Messina Seismological Observatory Memorial Volume

Edited by
Antonio Bottari
Dieter Mayer-Rosa
JesùsIbànez
Michèle Maugeri

2005

Birkhäuser Verlag
Basel • Boston • Berlin

Reprint from Pure and Applied Geophysics
(PAGEOPH), Volume 162 (2005) No. 4

Editors:

Antonio Bottari
Osservatorio Sismologico
Université di Messina
Via Osservatorio, 4
98121 Messina
Italy

Jesùs Ibànez
Istituto Andaluz de Geofisica
Universidad de Granada
Campus de Cartuja
18071 Granada
Spain

Dieter Mayer-Rosa
Institut für Geophysik
ETH Hönggerberg
CH-8093 Zürich
Switzerland

Michèle Maugeri
Dipartimento di Ingegneria Civile
ed ambientale
Université di Catania
Viale A. Doria, 6
95125 Catania
Italy

A CIP catalogue record for this book is available from the Library of Congress, Washington D.C., USA

Bibliographic information published by Die Deutsche Bibliothek:
Die Deutsche Bibliothek lists this publication in the Deutsche Nationalbibliographie; detailed bibliographic data is available in the internet at <http://dnb.ddb.de>

ISBN 3-7643-7263-X Birkhäuser Verlag, Basel - Boston - Berlin

© 2005 Birkhäuser Verlag, P.O.Box 133, CH-4010 Basel, Switzerland
Part of Springer Science+Business Media
Printed on acid-free paper produced from chlorine-free pulp
Printed in Germany

ISBN-10: 3-7643-7263-X
ISBN-13: 978-3-7643-7263-7

9 8 7 6 5 4 3 2 1

PURE AND APPLIED GEOPHYSICS
Vol. 162, No. 4, 2005

Contents

Pure appl. geophys. 162 (2005) 657
0033–4553/05/040657–1
DOI 10.1007/s00024-004-2629-8

▌Pure and Applied Geophysics

Messina Seismological Observatory Memorial Volume

Preface

The Seismological Observatory of Messina University, reformed in 1998 as a research structure aimed at the study of topics related to seismicity and seismic risk mitigation, took charge of the observation and study heritage of the Messina Seismological Observatory which, set up in 1887 in the Nautical Institute, was joined to the Royal Institute of Earth Physics with a new venue in Osservatorio Street, n. 4 in 1905.

Destroyed by the 1908 December 28th earthquake, the Observatory recommenced its activity again during the early 20s, and in 1947 it was annexed to the Geophysical and Geodetic Institute of Messina University which was closed down in 1994 after the formation of departments.

This volume features papers produced both by teachers who have contributed to the reformation of the Seismological Observatory, and many authors who, long-lastingly, have collaborated with the Seismological Observatory and sought to remember, through their research, the disastrous 1908 December 28th earthquake, whose 90th anniversary was commemorated in Messina with an international meeting.

The papers, prevalently focused on topics related to the seismicity of the Messina Straits area, propose new methodological approaches and results achieved over the years in the fields of Geophysics, Applied Seismology and Engineering Seismology.

This volume is dedicated to the victims of the 1908 December 28th earthquake and, particularly, to the researchers and technicians who contributed, with the rebuilding of the Observatory, to the resumption of seismological observation activity.

Messina, 2004 May 18th

The President
Prof. Antonio Bottari

Pure appl. geophys. 162 (2005) 659–670
0033–4553/05/040659–12
DOI 10.1007/s00024-004-2630-2

© Birkhäuser Verlag, Basel, 2005

❙ Pure and Applied Geophysics

A Methodological Approach for the Evaluation of Urban and Territorial Seismic Vulnerability

A. TERAMO[1], A. BOTTARI[1], D. TERMINI[1], and C. BOTTARI[1]

Abstract — A methodology of urban and territorial vulnerability analysis related both to engineering studies and social priority levels is presented. Modelling an urban system as a set of elements and relations, the evaluation of seismic vulnerability is carried out through suitable functions that characterize the seismic reliability of the site in terms of the geotechnical reliability of the ground and a structural reliability of buildings regarding the importance of activities referable to each element of the system. The results of the study, represented in map form, allow a full characterization of the distribution of urban and territorial vulnerability aimed at a seismic risk mitigation.

Key words: Seismic vulnerability, urban and territorial systems, safety reducers, social priority levels.

1. Introduction

The proposed methodological approach is aimed at the characterization of urban and territorial vulnerability that depicts itself as an operative and strategic instrument for planning not restricted to the urban context, but extended to the entire municipal or provincial territory, through a set of functions that can correlate different degrees of site seismic response to the different characteristics of the elements of the territory.

In particular, it is highlighted that the seismic vulnerability evaluation of an urban or territorial system, carried out through usual approaches (like damage scenarios depicted on the structural characteristics of buildings, microzonation studies, implications of macroeconomic vulnerability, earthquake loss estimate, etc. (FÄH *et al.*, 2001; CHEN *et al.*, 2001; AUGUSTI *et al.*, 1985, 2001; LEKKAS, 2002; BENEDETTI *et al.*, 1988; BRAGA *et al.*, 1987; GAVARINI and ANGELETTI, 1984; GAVARINI, 2001), do not correlate, in general, the safety reducers of the urban system elements to their level of social and strategic relevance.

---
[1] Osservatorio Sismologico, Università di Messina,Via Osservatorio, 4 - 98121 Messina, Italy.
E-mail: teramo@unime.it

The consequence is the actual impossibility for local administrations to plan, within a seismic prevention program, suitable interventions with priority levels related to sustainable costs.

The proposed approach allows the evaluation of urban and territorial seismic vulnerability, on a large scale, in relation to both specific site factors like the geotechnical reliability of the ground or the structural reliability of buildings and the social priority level of each element of the system within the territorial context. In such a way it is possible to depict seismic upgrading works of existing constructions whose level of vulnerability is not consistent with the location of strategic structures and emergency streets and areas. To this end the territorial system is modelled through a set of point, line or area elements, with a private and/or public character (such as buildings, streets and areas), and relations existing among these elements related to their importance within the urban system.

A significant test of the proposed procedure has been conducted with reference to the city of Melicucco located in Southern Calabria (Italy), because of its particular urban and territorial context and the high level of regional seismicity.

2. *The Evaluation Methodology of the Urban and Territorial Seismic Vulnerability*

The model of the urban and territorial system, adopted for the evaluation of seismic vulnerability, is characterized by a set of elements (streets, buildings and areas) and functions that model both their structural reliability and social relevance.

The seismic vulnerability function Λ, is defined in the set of a_i elements of the territory through two functions χ and ψ which model the local and territorial safety reducers (factors which reduce safety and are referable to the seismic reliability of each element of territory) and the social priority levels (referable to the importance of the single activities within the system), respectively. It can be written as:

$$\Lambda(a_i) = f[\chi(a_i) : \psi(a_i)] \tag{1}$$

Through these functions χ and ψ, that supplement the distribution of seismic vulnerability levels with specific factors referable to their strategic importance within the territory, it is possible to arrange suitable maps to plan interventions of seismic upgrading works on point, line and area elements of territory which, with respect to the sustainable costs, minimize the Λ function (1) with reference to the need levels of residual functionality of the urban system after an earthquake.

Therefore, taking into account the specific characters of the urban and territorial elements, the evaluation of the seismic vulnerability function is carried out through a set of equivalent distributions of vulnerability:

$$\Lambda^e(a_{ik}) = f[\chi(a_{ik}); \psi(a_{ik})], \tag{2}$$

where i individualizes the general element of territory a (as a point, line or area) to which the evaluation of vulnerability refers; k characterizes its type: $k = 1$ refers to the *morpho-typological characters of buildings* (structural typology and seismic reliability of buildings, unauthorized buildings, etc.); $k = 2$ refers to the *collective (or public) system* (distribution of services, productive assets and public buildings; etc.); $k = 3$ refers to the *critical spatial elements* (streets, safety routes, strategic structures, etc.).

As a consequence, the vulnerability level of a strategic building (hospital, school, aid coordination center) whose safety reducers have been calculated through surveys *in situ*, is characterized by a social priority level greater than a trading center or a house. And besides, the level of vulnerability of a motorway that links up, for instance, the urban center and the surrounding area with the entire region, which includes bridges and viaducts for which the safety reducers have been calculated, is characterized by a social priority level greater than that of a provincial, municipal or country road that links up a small number of houses.

So that, through suitable maps of equivalent distribution of seismic vulnerability, it is possible to characterize a high level of vulnerability of a given construction as the result of safety reducers of little (great) importance and a high (low) level of social priority.

For the Λ function in relation (1), an empirical relation can be assumed of the type:

$$\Lambda(a_{ik}) = \frac{1 + log\chi(a_{ik})}{1 - log\psi(a_{ik})} \tag{3}$$

The determination of the $\chi(a_{ik})$ function, carried out from case to case through specific and experimental type *in situ* surveys, allows a modelling of the *safety reducers* for the *i-th* element of territory with k typology a_{ik} (referable to the structural decay of buildings, lack of maintenance, faults in their construction or design, etc.), which determine a greater level of vulnerability. Four different levels of safety reducers are foreseen, equal to zero, low, average or high, respectively, to which four values of $\chi(a_{ik})$ function equal to 1, 2, 3 or 4 correspond.

The function $\psi(a_{ik})$ is determined from case to case modelling the *social (or strategic) priority levels* of the a_{ik} elements of territory, connected to the importance of the activities or the role within the system of each element to which, consequently, a greater level of relation density corresponds. Its values fall within the interval $0 < \psi \le 4$, with reference to the normalized social costs referable to the loss of a strategic structure, and can be evaluated as the ratio between the number of users of the given structure and the number of inhabitants of the city.

The diagram of the Λ function, shown in Figure 1, highlights the incidence of social priority levels in the determination of the vulnerability of the territory elements through the evaluation of the safety reducers. In this regard, it is to be noticed how a reduced level of social importance of a construction (an empty house located in the country, to which a social priority level close to zero corresponds) with difficult

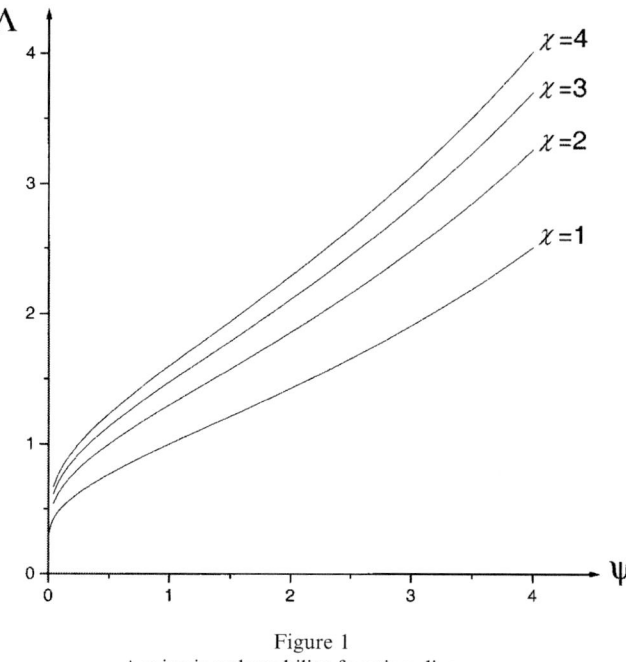

Figure 1
Λ seismic vulnerability function diagram.

stability conditions (for a safety reducer level equal to 4), determines a low level of seismic vulnerability.

3. Remarks on the χ and ψ Functions

A complete characterization of the χ function for the structures whose level of vulnerability has to be determined is related to the seismic reliability of the elements of the system. This system involves the ground stability with respect to the seismic shaking and the safety reducers of a building, a street or a bridge not consistent with the actual level of seismic stress due to the geotechnical and seismic characteristics of the ground.

It requires detailed knowledge of the geological, geophysical and geotechnical characteristics of the studied area through specific surveys.

The application of a quick seismic microzonation procedure (TERAMO *et al.*, 2005), at low cost, aimed at the characterization of different levels of seismic response of the ground, in terms of γ site amplification factors of expected peak ground acceleration, allows an evaluation of seismic reliability of the structures investigated. This procedure, which corresponds to a level between 2 and 3 of seismic microzonation described by the Technical Committee for Earthquake Geotechnical

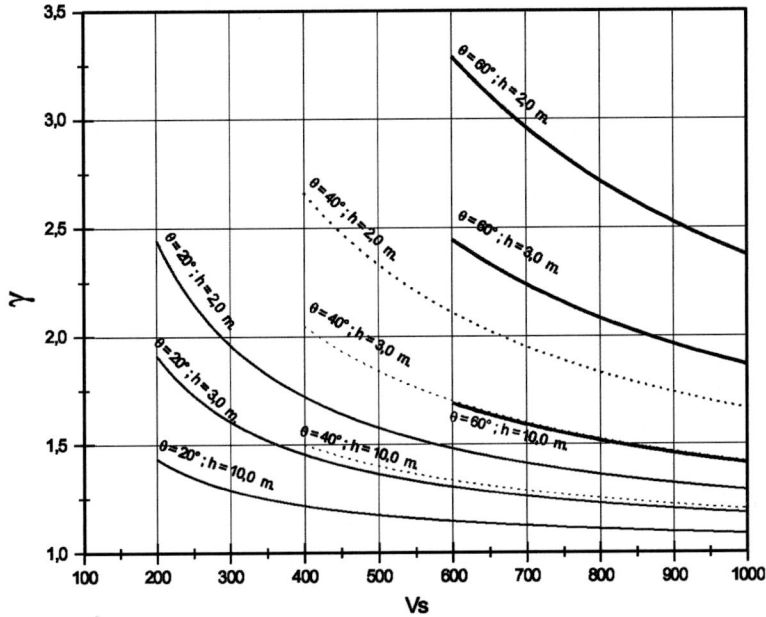

Figure 2
Site amplification factors for a P.G.A. 0.25 g.

Engineering of the International Society for Soil Mechanics and Foundation Engineering (T.C.4, 1993), correlates the incidence of the geological conditions on the change of the local seismic response of the ground through four different geo-referenced layers of each micro-zone in which the whole territory is subdivided, and referable to: the slope of soils, maximum expected acceleration a on bedrock, shear-wave velocities of surface soils and depth of the ground water table h.

For each micro-zone the site amplification factors γ are calculated through the following empirical relation (TERAMO et al., 2005):

$$\gamma = 1 + \alpha \frac{1n^{3.7}a}{1nh} \qquad (4)$$

being α a coefficient depending on slope soil and shear-wave velocities.

To make the incidence of each variable on the site amplification factors clearer, the diagram of Figure 2 is shown, in which a fixed level of expected acceleration on bedrock ($a = 0.25$ g) is associated with different values of slope ($\theta = 20$ °, 40°, 60°) and various depths of the surface ground water table ($h = 2$ m, 3 m, 10 m).

The determination of the values of χ function for each element of the territory is made through a quick approach or an exact one with respect to the type of vulnerability survey related to the size of the area or the number of constructions to be investigated.

In the first case, values falling in the interval [0,4] will be adopted in relation to the level of safety reducers (low, medium, high, very high) depicted for the constructions, owing to the values of site amplification factors γ or their structural reliability. In the second, an exact determination of the values of χ is made through both specific dynamic type surveys *in situ* and a modal testing on the elements of the existing structures to individualize eventual pathologies referable to as a lack of maintenance or faults in construction or design, and taking into account, through a suitable computer program, the effects on the structure of the site amplification factors of the expected acceleration. Values of χ function falling in the interval [0,4] will be adopted as result of a comparison between actual behavior and theoretical behavior of a building.

It is to be noted that the proposed procedure allows the limits of applicability of the usual ones to be overcome, evaluating the influence of seismic upgrading works on the vulnerability of the whole construction, to reach a prefixed vulnerability level consistent with the location of strategic roads and/or escape routes. From all this derives a set of constructions with a prefixed vulnerability level that represents the basis of an efficient program of seismic risk mitigation.

Finally, the determination of the ψ function that models the *social priority levels* of the territorial and urban system, is formed evaluating the social losses in monetary terms through a macroeconomic indicator that can give a suitable characterization of the importance or the social relevance of a structure at municipal, district or regional level, taking into account the presence of similar structures in the territory. The social costs, referable to the loss of a productive structure or a sanitary structure strictly connected to the economy of the city or the district, must be evaluated, from case to case, correlating the total number of users of each structure to the number of inhabitants of the city.

4. An Application of the Proposed Approach

An application of the proposed procedure has been formulated out in the city of Melicucco in Southern Calabria, Italy. This city falls within the n. 69 seismogenic zone in the seismic zonation (SCANDONE *et al.*, 1992) adopted by GNDT (Gruppo Nazionale per la Difesa dai Terremoti, C.N.R., Rome) for the seismic hazard of Italy, in the last version of April 1996 (Fig. 3); it is located in the Gioia Tauro plain, in a district of 33 cities whose territories move within the orbit of the great commercial port of Gioia Tauro, an important economic area of the Mediterranean Sea. The main economic assets of the city of Melicucco which, with about 5000 inhabitants and a population density of 827,40 per km^2 shows the greatest population increase of the cities in the district at the last census, are correlated to the agricultural, artisan and tourist assets that have produced, over the years, a

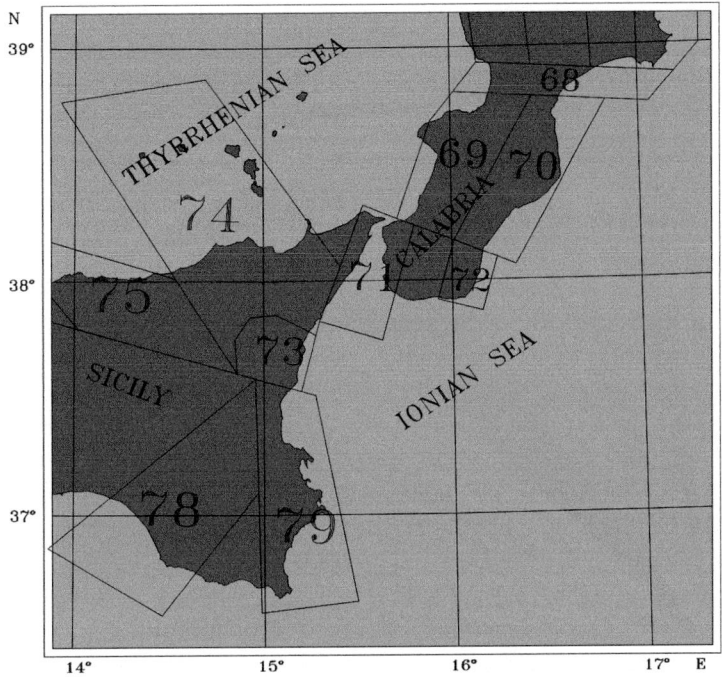

Figure 3
Seismogenic zonation of Southern Italy (SCANDONE et al., 1992).

progressive development of the original structure of the city, well connected to the motorways and other cities of the district.

The building estate consists of about 1900 houses, 37% of which fall within the historical center, 19% of which were built before 1945, 40% between 1946 and 1971, and 41% after 1972. The more recent buildings are mainly reinforced concrete structures with 3 or fewer floors; the others are one or two story stone or brick buildings.

The level of seismic hazard of the Gioia Tauro plain, within which the city of Melicucco lies, is characterized by an average PGA value equal to 0,386 g corresponding to a return period equal to 475 years (GNDT - CNR, 1996). In fact, this area is among those at greatest seismic risk with maximum observed intensities on this site attributable to earthquakes which occurred in Calabria and Sicily: East Sicily, 02–04–1169 (VI MCS); East Sicily, 11–01–1693 (V-VI MCS); Taurianova (Calabria) 28–03–1783 (XI MCS); Serre (Calabria), 05–11–1659 (VII-VIII MCS); Reggio Calabria, 16–11–1894 (VII-VIII MCS); the Messina Straits, 28–12–1908 (VIII MCS); Vibo (Calabria), 07–03–1928 (VII-VIII MCS).

The site amplification factors of the expected PGA used for the seismic reliability check of structures, have been evaluated with relation (4) through parameters deduced from specific *in situ* surveys aimed at the determination of shear-wave velocities of the surface soils (refraction seismic profiles), the depth of the ground water table (hydrogeological studies or data relative to surveys *in situ*), the slope of soils (obtained through a GIS software with a digital model of soil). It is to be observed that, within some areas of the urban context, high amplification factors of expected acceleration on bedrock ($\gamma \leq 2,5$), due to lithologies with a low level of shear-wave velocities or a low depth of the ground water table, have been determined.

The *safety reducers* modelled from the $\chi(a_{ik})$ function of a building ($k = 1$ typology) are essentially due to an absence or lack of maintenance, structural decay, faults in construction or design, unauthorized structural modifications, inadequate dimensions of structures. The highest values of χ (equal to 3 or 4) have been determined with greater density within the area of the historical center (due to great number of buildings structurally lowered) with respect to the rest of the city, where high values of χ have been evaluated only in correspondence with unauthorized buildings. For the $k = 2$ typology structures (productive assets, public buildings, services, etc.) located, for the most part outside the historical center, the χ function has rather low values (equal to 1 or 2), considering that they deal with recent structures not subjected to upgrading works or structural modifications.

The safety reducers of streets, safety routes, strategic structures with typology $k = 3$ are essentially due to the small width of the streets, the ground instability on which important roads were built, and due to low seismic reliability of some buildings that overlook small streets. The streets in the historical center, that must guarantee access to emergency areas located outside are characterized by high values of χ function (equal to 3 or 4).

The *social priority levels* of the system, modelled from the function $\psi(a_{ik})$, through local macroeconomic indicators related to the importance of the activities carried out in the city, essentially agricultural, artisan and tourist types have been evaluated. To this end, normalized values of ψ function have been determined for each typology of structures, through data relative to the number of users with respect to the number of inhabitants in the city of Melicucco.

Using GIS software, the equivalent distribution of the levels of seismic vulnerability, characterized through the function $\Lambda^e(a_{ik})$ of the relation (3), have been drawn up with reference to each k typology of the elements of territory. In particular, only for the $k = 1$ (*morpho-typlogical characters of buildings*) and $k = 2$ (*collective system*) typologies, minimum values of ψ or χ functions have been considered, respectively, to highlight the distribution of the correspondent levels of seismic vulnerability better.

In Figures 4, 5 and 6 extracts of these distributions of the functions $\Lambda^e(a_{ik})$ of the municipal territory of Melicucco are shown. Darker colors indicate a greater level of vulnerability. Note that higher levels of seismic vulnerability can be observed in the historical center (Fig. 4), because of the great number of old buildings. In Figures 5

Figure 4
Equivalent seismic vulnerability related to the morpho-typological characters of buildings.

Figure 5
Equivalent seismic vulnerability related to the collective system.

and 6 the distribution of seismic vulnerability levels related to the distribution of services, productive assets, etc. , as well as streets and safety routes for reaching emergency areas and strategic buildings are shown, respectively.

This type of representation allows, moreover, a full characterization of the distribution of the seismic vulnerability levels (Fig. 7), determined with the $\Lambda(a_{ik})$ functions, or a selective one (Figures 4, 5 and 6) with the equivalent vulnerability functions $\Lambda^e(a_{ik})$. Using maps of this type it is possible to make comparisons

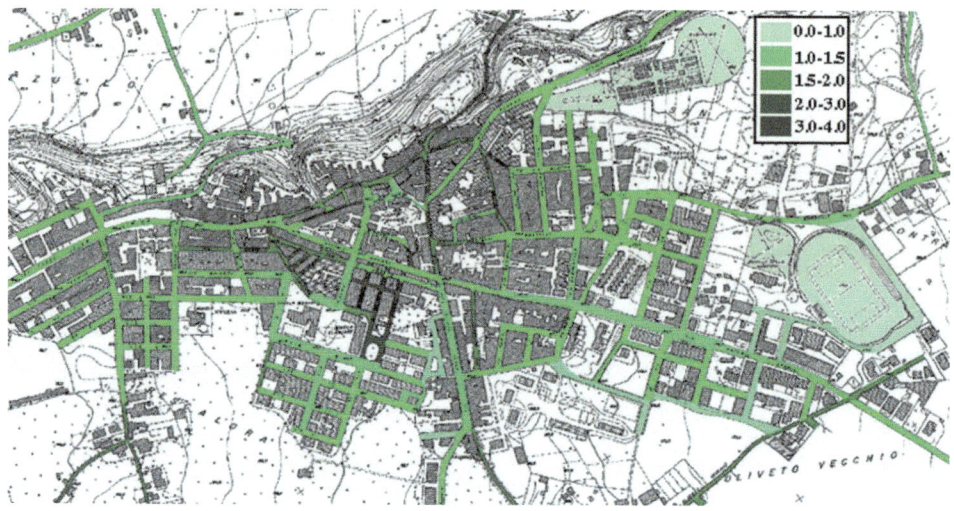

Figure 6
Equivalent seismic vulnerability related to the critical spatial elements.

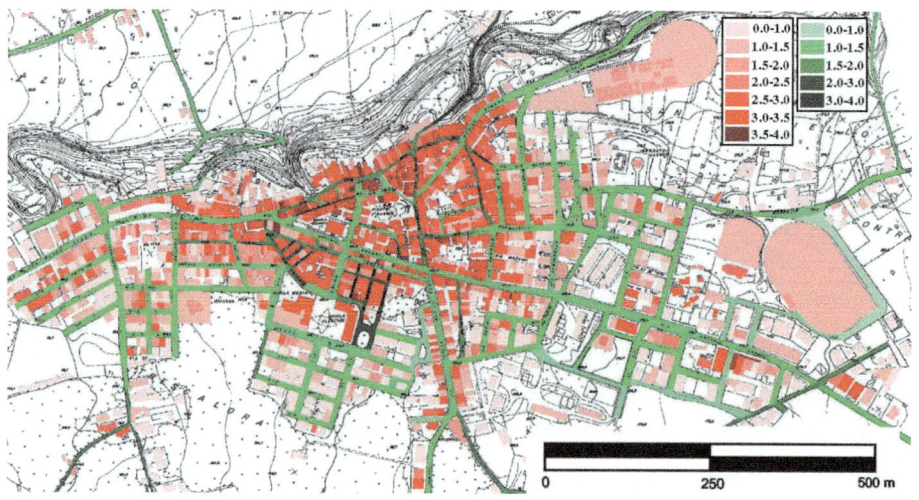

Figure 7
Seismic vulnerability distribution.

between several distributions of vulnerability for the individualization of threshold values and consequently, to plan seismic upgrading works of existing constructions, with respect to the sustainable costs, with the aim of checking the actual residual functionality levels of the system after an earthquake.

5. Conclusions

The proposed methodological approach for the evaluation of seismic vulnerability of a given urban system moves from the need to depict an operative tool for the characterization of an efficient strategy of seismic risk mitigation. It is strictly connected both with the evaluation of seismic hazard of the area in study and specific indicators suitable for representing different levels of vulnerability of the system. Such indicators have been characterized, modelling the system as a set of elements and relations, through specific functions that model the so-called safety reducers and social priority levels of the system: the first ones, relative to the seismic reliability of structures, consider eventual pathologies due to a structural decay or to a lack of maintenance or faults in construction or design, with reference to the seismic response of the ground obtained through the application of a quick seismic microzonation; the second takes into account the importance of some services and productive assets within the territorial system, in terms of social costs regarding the loss of such structures.

Employing these indicators it is possible to arrange different types of maps which correspond to the different typologies of the elements of territory that produce a spatial equivalent distribution seismic vulnerability, related to the morpho-typological characters of buildings, the collective (public) system and the critical spatial elements, respectively. A full or selective representation of the seismic vulnerability distribution of the territorial system is obtained relative to the number and type of adopted indicators.

The most significant result is the arrangement of maps of seismic vulnerability distribution related both to the actual conditions of buildings in the city and the importance of activities within the system. A comparison of different scenarios of vulnerability allows interventions in an upgrading process of existing constructions to be planned, aimed at a seismic risk mitigation.

REFERENCES

AUGUSTI, G., BENEDETTI, D., and CORSANEGO, A. (1985), *Investigation on seismic risk and seismic vulnerability in Italy.* In Proc. 4th Internat. Conf. *On Structural Safety and Reliability* (ICOSARR '85), Kobe, Japan, vol. 2, pp. 267–276.

AUGUSTI, G., CIAMPOLI, M., and GIOVENALE, P. (2001), *Seismic Vulnerability of Monumental Buildings,* Structural Safety *23*, 253–274.

BENEDETTI, D., BENZONI, G.M., and PARISI, M.A. (1988), *Seismic Vulnerability and Risk Evaluation for Old Urban Nuclei.* Earthq. Engin. Struct. Dyn. *16*, 183.

BRAGA, F., DOLCE, M., and LIBERATORE, D. (1987), *Rassegna critica dei metodi per la stima della vulnerabilità.* In Atti 3° Conv. Naz. L'ingegneria Sismica in Italia, Roma, 1987.

CHEN, Y., CHEN, Q., and CHEN, L. (2001), *Vulnerability Analysis in Earthquake Loss Estimate,* Natural Hazards *23*, 349–364.

FÄH, D., KIND, F., LANG, K., and GIARDINI, D. (2001), *Earthquake Scenarios for the City of Basel,* Soil Dyn. Earthq. Engin. *21*, 405–413.

GAVARINI, C. (2001), *Seismic Risk in Historical Centers,* Soil Dyn. Earthq. Engin. *21,* 459–466.

GAVARINI, C. and ANGELETTI, P. (1984), *Assessing seismic vulnerability in view of developing cost/benefit ratios for existing r.c. buildings in Italy.* In Proc. 8th World Conf. *On Earthquake Engineering,* San Francisco, vol. 1, 445.

GNDT (CNR) (1996), *Pericolosità sismica del territorio Nazionale.* http://emidius.itim.mi.cnr.it/GNDT/PS.html

LEKKAS, E.L. (2002), *The Role of Earthquake-related Effects in Urban Complexes,* Natural Hazards *25,* 23–35.

SCANDONE, P., PATACCA, E., MELETTI, C., BELLATALLA, M., PERILLI, N., and SANTINI, U. (1992), *Struttura Geologica, evoluzione cinematica e schema sismotettonico della Penisola Italiana.* Atti del convegno GNDT, Pisa 25–27 Giugno 1990, 1, 119–135. Edizione Ambiente, Bologna.

TC4 (1993), *Manual for Zonation on Seismic Geotechnical Hazards.* Technical Committee for Earthquake Geotechnical Engineering, TC4, of I.S.S.M.F.E. Ed. The Japanese Society of Soil Mechanics and Foundation Engineering. December 1993. 149 pp.

TERAMO, A., MAUGERI, M., BOTTARI, A., TERMINI, D., and BOTTARI, C. (2005), *A Quick Seismic Microzonation of Wide Areas,* Pure Appl. Geophys, *162,* 671–682

(Received January 20, 2004, accepted March 22, 2004)

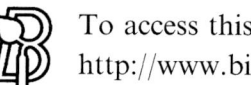

To access this journal online:
http://www.birkhauser.ch

Pure appl. geophys. 162 (2005) 671–682
0033–4553/05/040671–12
DOI 10.1007/s00024-004-2631-1

▌Pure and Applied Geophysics

On a Quick Seismic Microzonation of Wide Areas

A. Teramo[1], M. Maugeri[2], A. Bottari[1], and D. Termini[1]

Abstract—In this paper a characterization of seismic site response is proposed, taking into consideration geo-morphological conditions, geotechnical and geophysical parameters such as slope, average shear-wave velocity, maximum expected acceleration on bedrock, depth of ground water table. An empirical relationship is presented between these parameters and applied, with the objective of determining ground motion amplification coefficients to be used in specific programs of land use or town planning dedicated to the mitigation of seismic risk.

Key words: Site effects, quick seismic microzonation, seismic risk, ground motion amplification.

1. Introduction

The development of a procedure of quick microzonation of wide areas (from small parts of municipal territory to provincial and regional ones) is motivated by the need to have a quick and low cost instrument for the evaluation of the actual level of urban and territorial seismic risk regarding the seismic hazard and geological and geotechnical characteristics of study areas.

The application of preventive seismic microzonation to a wide area, before the occurrence of a strong seismic event, is surely to be preferred to a comparison between the seismological, geological, and geotechnical data of the same area and the effects evaluated after the earthquake. The procedure explained as follows has been chosen to allow reliable development of a safe territorial design with the main objective being the contribution to the mitigation of seismic risk.

The implementation of such a procedure initially requires an execution of territorial surveys. The level of seismic hazard in the area is determined by starting with the statistical analysis of the observed distribution of seismicity in the seismogenic zones within which these areas fall. The possible occurrence of local instability due to slope failure and/or low depth ground water table is taken into

[1] Osservatorio Sismologico, Università di Messina, Via Osservatorio, 4 - 98121 Messina, Italy. E-mail: teramo@unime.it
[2] Dipartimento di Ingegneria civile e ambientale, Università di Catania Via A. Doria, 6 - 95125 Catania, Italy.

consideration also through specific geological, geophysical and geotechnical surveys extended to the whole study area.

The procedure corresponds to a level between 2 and 3 of definition of seismic microzonation described by the Technical Committee for Earthquake Geotechnical Engineering of the International Society for Soil Mechanics and Foundation Engineering (T.C.4, 1993). In comparison with the usual procedures, it offers clear advantages of simplicity due to the adoption of only one parameter that correlates prefixed seismic hazard levels to local geomorphological conditions. This parameter distinguishes itself as a suitable site amplification factor of the peak ground acceleration.

The procedure was applied in southern Italy, the entire provincial territory of Siracusa City, Sicily, and the territories of the cities of Melicucco and Rosarno, Calabria (Fig. 1), both areas being characterized by strong earthquakes (M > 7). Because of the complexity of structural and geomorphologic characteristics, difficult problems in seismic planning and administration must be expected with regard to the destructive effects on buildings and the environment.

For the characterization of site amplification factors we considered the results of many studies carried out by various researchers on strong earthquakes which have

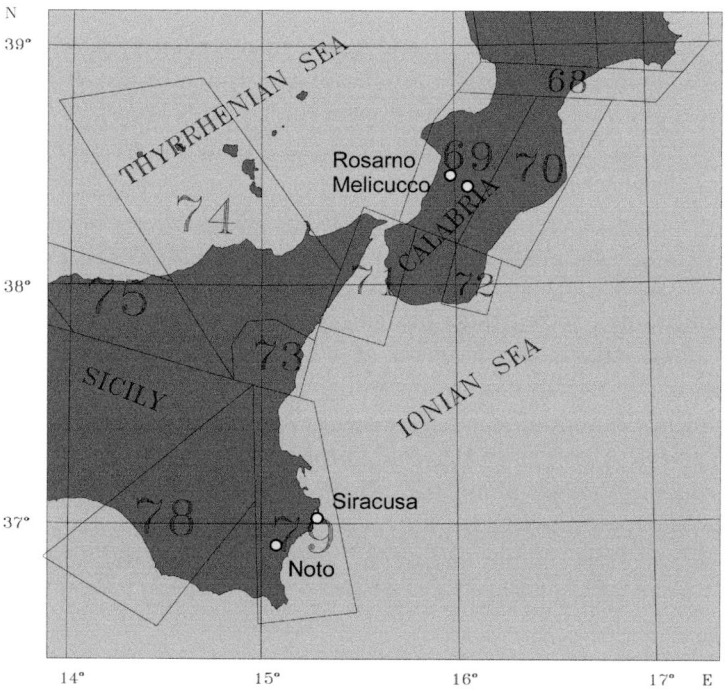

Figure 1
Seismogenic zonation of southern Italy (SCANDONE *et al.*, 1992).

occurred in Italy over the years (FACCIOLI and AGALBATO, 1979; FRENNA and MAUGERI, 1994; MAUGERI and CARRUBBA, 1985; MARCELLINI *et al.*, 2001; MASSIMINO *et al.*, 2001). Particular reference was taken for the studies conducted in the areas of southeastern Sicily (around the cities of Noto and Siracusa) falling within seismogenic zones 78 and 79 (Fig. 1) of the seismic zonation of the Italian territory (SCANDONE *et al.*, 1992). They are characterized by an expected peak ground acceleration a between 0.24 g and 0.28 g, for a return period of $T = 475$ years (GNDT – CNR, 1996).

2. The Field Surveys

The determination of lithotypes of soils was the first level of survey for the acquisition of the data used for the characterization of local surface layers, which are of greater interest from a geotechnical point of view. To this end, geological maps were collected and supplemented by additional data taken from literature and checked *in situ*. However, special surveys had to be produced for sites where the individual lithostratigraphycal characteristics are particularly complex.

In order to evaluate the effect of particular geological conditions on the change of the local seismic response of ground motion, geophysical surveys were performed for the lithotypes with a significant extent in the territory. Using the results obtained in different geophysical measurements, average shear-wave velocities of surface soils in the uppermost 30 meters (EUROCODE EC8, 1974) of soil were determined.

Previous studies on ground motion behavior (MEDVEDEV, 1962; JOYNER and FUMAL, 1988; BORCHERDT *et al.*, 1991) have revealed a correlation between this average shear-wave velocity and the site amplification factor. Additionally in the present paper, the dependence on local soil instability conditions related to mechanical characteristics of surface soils, such as the slope of soils and the depth of ground water table, was taken into account. The first one was determined using suitable GIS software (ESRI, Arcview) with a DTM (Digital Territory Model) with a 20-m grid resolution; the second was determined using data taken from literature or surveys *in situ*. Both parameters refer to a significant reduction of the cohesiveness and friction resistance of surface soils leading to a remarkable reduction of the geotechnical stability of soils affecting the construction of civil and industrial buildings.

3. The Characterization of the Seismic Site Response

The main goal of the proposed procedure is to determine, for each defined zone of the study area, an amplification factor of peak ground acceleration (PGA) that can be used to depict different levels of soil stability.

The study area, in this case the entire provincial territory of the city of Siracusa, has been subdivided into numerous microzones by a 20-m grid. Four different georeferenced layers of these microzones were adopted; one for each parameter used for the characterization of the seismic response of the ground: slope, acceleration, shear-wave velocity and depth of ground water table.

For each microzone the site amplification factor γ has been calculated through an empirical relation in the following form:

$$\gamma = \Gamma(\theta, a, 1/v_s, 1/h),$$

where $\theta =$ slope of soils, $a =$ peak ground acceleration, $v_s =$ average shear-wave velocities of surface soils and $h =$ depth of ground water table. With regard to the maximum expected acceleration a on bedrock, it is necessary to point out that the proposed relation is valid with respect to strong earthquakes ($M > 6$) and, therefore, to a rather nonlinear behavior of ground.

In this relationship, the proportionality levels of the site amplification factor with θ, v_s and h have been depicted according to the increases suggested by different rules and regulations (RèGLES PS 92, 1992; TC4, 1993; EUROCODE EC8, 1994;) or adopted in specific studies (MEDVEDEV, S.V., 1960), or grade-3 microzonations. These microzonations refer to MAUGERI and FRENNA (1995) for Augusta City, damaged by the 1990 southeastern Sicily earthquake ($M = 5.6$); MAUGERI and CARRUBBA (1997) for Calabritto City, damaged by the 1980 Irpinia earthquake ($M = 6.6$); TENTO *et al.* (2001) for Fabriano City and CAPILLERI *et al.* (2001) for Sellano City, both damaged

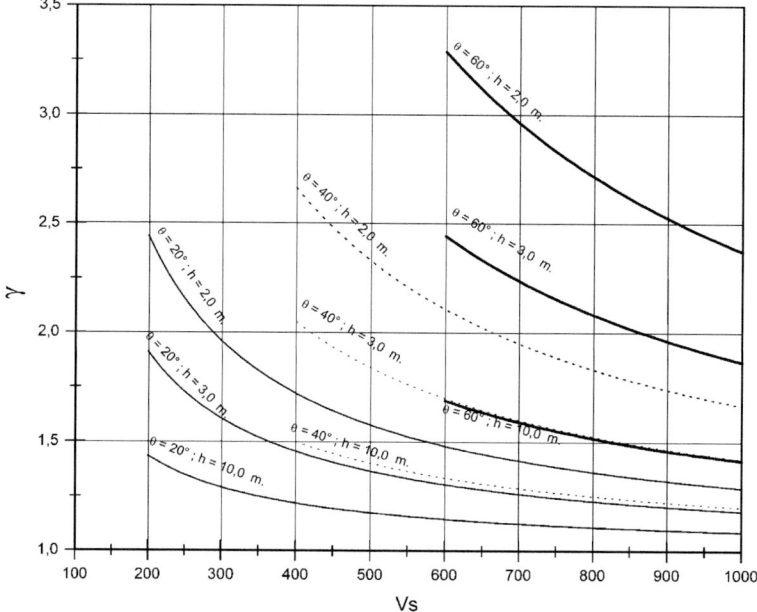

Figure 2
Site amplification factors for a P.G.A. 0.25 g.

by the 1996–97 Umbria-Marche earthquake (M = 5.8); MASSIMINO and MAUGERI (2001) for Noto City, for the 1693 scenario earthquake (M = 7.3).

In fact, from the rules and regulations mentioned above it is possible to derive several amplification factors relative to different values of soil slope θ (1.0 for $\theta \leq 15°$, to 1.4 for $\theta \geq 30°$), of shear-wave velocity v_s (1.0 for $v_s = 800$ m/s, to 3 for $v_s = 200$ m/s), and depth of the ground water table h (1.0 for $h = 10$ m, to 1.4 for $h = 2.0$ m), respectively.

The relation proposed here is:

$$\gamma = 1 + \alpha \frac{\ln^{3.7} a}{\ln h}$$

being $\alpha = (1 + \mathrm{tg}\theta)/2v_s$ and where a, v_s and h are given in [cm/sec²], [m/sec] and [m], respectively) and correlates all parameters that influence the site amplification factors. Their values are consistent with the above-mentioned rules, regulations and studies of seismic microzonation, and allow the set-up of a quick seismic microzonation.

In order to demonstrate the effect of each variable on the site amplification factor better, some diagrams are shown (Figs. 2, 3, 4), in which different levels of expected acceleration on bedrock ($a = 0.25$ g; 0.10 g; 0.05 g) are associated with different values of slope ($\theta = 20°$, 40°, 60°) and various depths of surface ground water table ($h = 2$ m, 3 m, 10 m.).

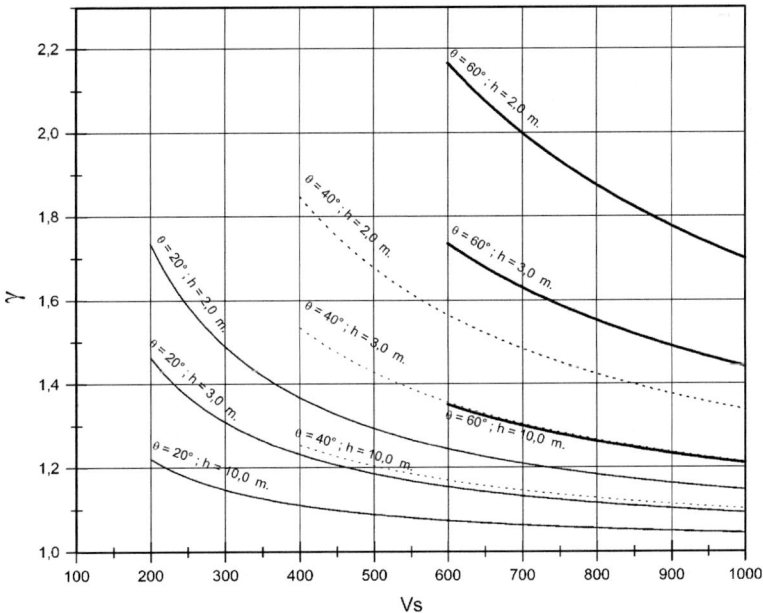

Figure 3
Site amplification factors for a P.G.A. 0.10 g.

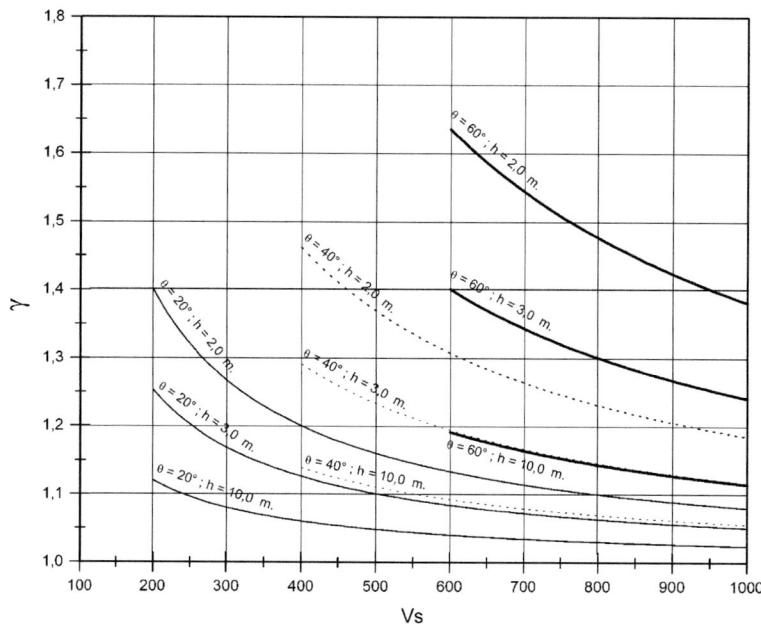

Figure 4
Site amplification factors for a P.G.A. 0.05 g.

All curves have been plotted for v_s values consistent with the corresponding levels of slope. It is assumed that an alluvium with shear-wave velocity ≤ 200 m/s has a slope ≤ 20° and, similarly, a rock with a slope ≥ 60 °, will have a value of v_s ≥ 600 m/s. In any case, through the proposed relation, it is possible to determine the amplification factor corresponding to the actual values of the considered parameters.

The variability of the site amplification factor with the expected acceleration at bedrock level is shown in Figure 5, with reference to a fixed soil slope value ($\theta = 40$ °) and shear-wave velocity ($v_s = 400$ m/s), and for different values and depth h of ground water table. Such relationships are confirmed by microzonation studies carried out by SHIMA (1978) and MORA and VAHRSON (1994).

4. The Application of the Quick Seismic Microzonation

An application of such a procedure was performed in the entire provincial territory of Siracusa City (2180 km^2), subdivided into 8 sectors (Fig. 6). Each of them was further subdivided by a 20-m grid to allow greater detail in mapping the site amplification factors with respect to the local geomorphological, geophysical and geotechnical conditions.

The map of quick seismic microzonation (Fig. 7) was designed in scale 1:25.000 by grouping several tables of the IGM (Military Geographic Institute), compiling the

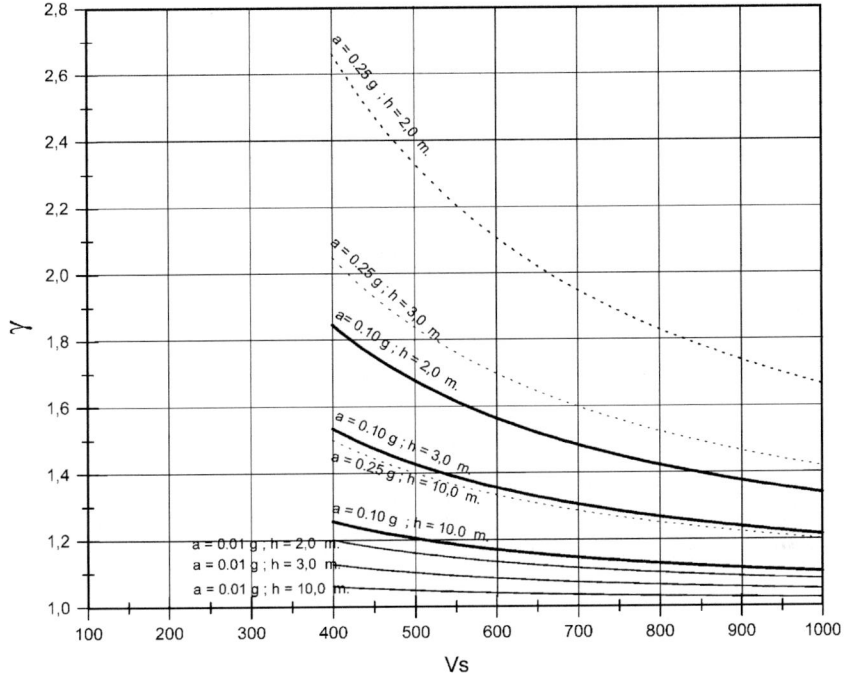

Figure 5
Site amplification factors for a fixed value of soil slope ($\theta = 40$ °).

above-mentioned field surveys, supplemented by data taken from literature and checked *in situ*.

In particular, shear-wave velocity values v_s between 200 and 1000 m/s, were adopted on the basis of refraction seismic profiles, or, in some cases, estimated with an error equal to ± 50 m/s. In the test site of the City of Noto, the shear-wave velocity was determined with cross-hole and down-hole geophysical surveys.

Depth values h of the ground water table, between 1 and 10 m, considering that for $h > 10$ m the effect on the site amplification is very small, were taken from a hydrogeological study carried out by Catania University - CNR (AURELI *et al.*, 1989).

The slope values were obtained through a GIS software, with reference to a grid size of 20 meters. The peak ground acceleration adopted is 0.25 *g*, according to the Italian Territory maps of seismic hazard (GNDT - CNR, 1996).

The validation of the quick seismic microzonation was accomplished through the above-mentioned grade-3 seismic microzonation of Noto City (MASSIMINO and MAUGERI, 2001). It is based on synthetic accelerograms obtained by modelling the Ibleo-Maltese fault failure that caused the 10-01-1693 earthquake, adopted as a scenario earthquake. The synthetic accelerograms, determined with reference to the

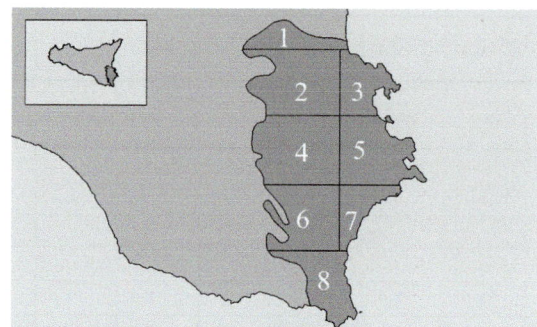

Figure 6
Subdivision of the whole provincial territory of Siracusa City in 8 sectors for the quick seismic
microzonation.

bedrock, were corrected for the surface level through a nonlinear calculation code.
The geotechnical characterization of the soil has been produced through tests *in situ*,
bore-hole, cross-hole and down-hole surveys, laboratory tests including resonant
column tests, cyclic torsional shear tests and cyclic triaxial tests (CAVALLARO *et al.*,
1999, 2002, 2003; MAUGERI and Cavallaro 2001).

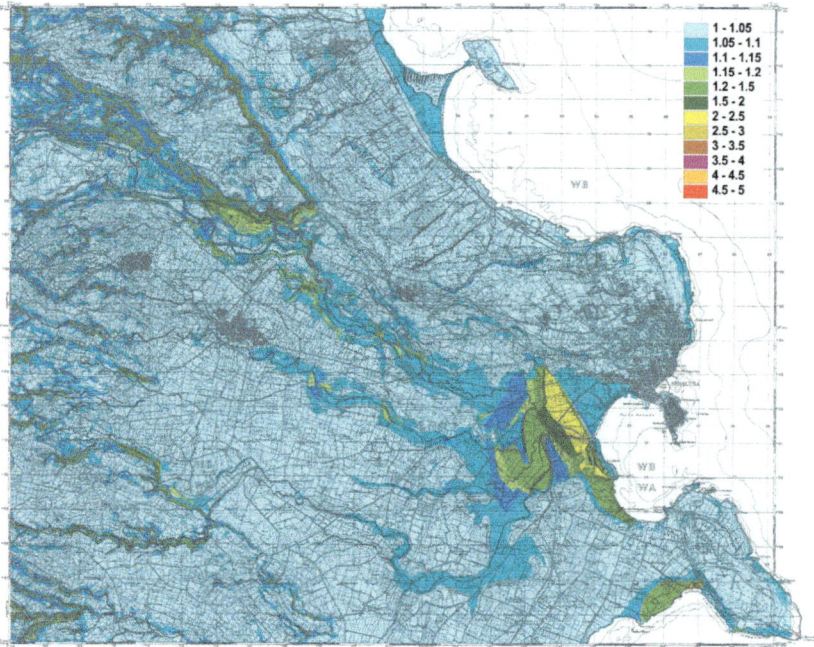

Figure 7
Quick seismic microzonation of the sector no. 5 of the provincial territory of Siracusa City (1:25000).

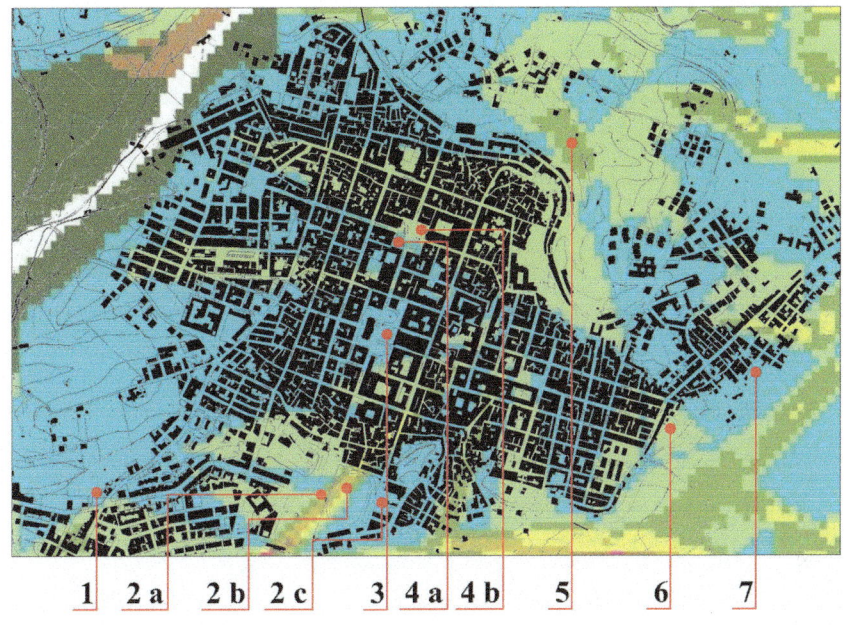

Figure 8
Quick seismic microzonation of Noto City: Location of sites for the comparison of the site amplification factors of the grade-3 seismic microzonation (MASSIMINO and MAUGERI, 2001) and the proposed quick one (1:4000)

In Figure 7 the quick seismic microzonation of sector no. 5 of the provincial territory of Siracusa City is shown. Different zones, to which different values of the site amplification factor are associated, are colour-coded from 1 to 5. However, only factors from 1 to 2.5 are applicable in this sector.

The reliability of the proposed quick seismic microzonation can be deduced through a comparison between the site amplification factors calculated with the proposed relation, and those taken from the grade-3 seismic microzonation of Noto City (Table 1), for different sites located in the centre of Noto City (Fig. 8) with greater seismic risk. A Komolgorov-Smirnov test shows, in fact, a significance of 5%. Moreover, it is to be highlighted that the uncertainties relative to the determination of shear-wave velocity values v_s of surface soils and depth values h of the ground water table, evaluated in about 10%, involve an error on the site amplification factor of about 7%.

The consistency level of the obtained results, considering different study scales (1:4000 for the grade-3 seismic microzonation and 1:25000 for the quick one), is remarkable considering the different detail level of the surveys carried out, respectively.

Table 1

Comparison of site amplification factors obtained through the grade-3 seismic microzonation by MASSIMINO and MAUGERI (2001) and the quick one at different sites of the center of Noto City.

Sites	Site ampl. factors (Grade-3 s.m.)	Site ampl. factors (Quick s.m.)
1	1.11	1.09
2a	1.28	1.25
2b	1.43	1.34
2c	1.11	1.09
3	1.17	1.09
4a	1.28	1.09
4b	1.17	1.17
5	1.43	1.25
6	1.11	1.17
7	1.11	1.09

4. Summary

An empirical relation for the determination of the site amplification factor of peak ground acceleration, related to the geological and geotechnical characteristics of the areas under study, is proposed with the aim of achieving a grade-2 seismic microzonation.

Amplification factors, starting with rules and regulations taken from literature, have been deduced through the proposed relation taking into account θ (soil slope), a (maximum expected acceleration on bedrock), v_s (average shear-wave velocities of surface soils) and the h (depth of the ground water table). Their consistency with several grade-3 microzonation studies carried out in different seismogenic zones, is evaluated.

The proposed approach has been checked in wide areas in southern Calabria and in southeastern Sicily. In the provincial territory of Siracusa City and in the center of Noto City, previous grade-3 seismic microzonation data were used for the validation of the procedure. The results of this study show a remarkable consistency with higher level studies and can be used for quick and efficient microzonation procedures.

REFERENCES

AURELI, A., ADORNI, G., CHIAVETTA, A.F., FAZIO, F., TAZZINA, S., and MESSINEO, G. (1989), *Carta della vulnerabilità delle falde idriche Settore Nord-orientale Ibleo. (Sicilia S.E.).* Università di Catania – CNR, Gruppo Nazionale per la difesa dalle catastrofi idrogeologiche.

BORCHERDT, R.D., WENTWORTH, C.M., JANSSEN, A., FUMAL, T., and GIBBS, J. (1991), *Methodology for predictive GIS mapping for special study zones for strong ground shaking in the San Francisco Bay Region*, Proc. Fourth International Conf. *Seismic Zonation, 3*, pp. 545–552.

CAVALLARO, A., LO PRESTI, D. C. F., MAUGERI, M. and PALLARA O. (1999), *A case study (The Saint Nicolò Cathedral) for dynamic characterization of soil from in situ and laboratory tests*, Proc. Second Internat. Symp. *Earthquake Resistant Engineering Structures*, Catania, 15–17 June 1999, pp. 769–778.

CAVALLARO, A. and MAUGERI, M. (2002), *Site characterization by in situ and laboratory tests at Noto*, Proc. Symp. *Geotechnical Analisys of Seismic Vulnerability of Historical Monuments*, Catania, 15 November 2001.

CAVALLARO, A. and MAUGERI, M. (2003), *Small strain stiffness from in situ and laboratory tests for the City of Noto soil*; Proc. Third International Symp. *Deformation Characteristics of Geomaterials*, Lyon, 22–24 September 2003.

CAPILLERI, P., MASSIMINO, M.R., and MAUGERI, M. (2001), *The Grade-3 Microzonation of Sellano*. Special issue: *The 1997-98 Umbria-Marche Earthquake*, Italian Geotechnical Journal, *XXXV*, n. 4.

EUROCODE EC8 (1994), *Design Provisions for Earthquake Resistance of Structures*, CEN, 1994, November.

FACCIOLI, E., and AGALBATO, D. (1979), *Attenuation of strong motion parameters during the 1976 Friuli earthquakes*, Proc. 2nd USA National Con. *Earthquake Engineering*, Stanford, California, pp. 233–242.

FINN, W.D.L. (1991), *Geotechnical engineering aspects of microzonation*, Proc. Fourth International Conf. *Seismic Zonation*. *1*, pp. 199–259.

FRENNA, S.M., and MAUGERI, M. (1994), *Microzoning of the Saline site of Augusta town (Sicily, Italy)*, Proc. 10th European Conf. *Earthquake Engineering*, Vien, August 28 – September 2, *1*, pp. 43–49.

GNDT (CNR) (1996), *Pericolosità sismica del territorio Nazionale*. http://emidius.itim.mi.cnr.it/GNDT/PS.html

JOYNER, W.B., and FUMAL, T. (1984), *Use of measured shear-wave velocity for predicting geological site effect on strong motion*, Proc. Eighth World Conf. *Earthquake Eng.*. *2*, pp. 777–783.

MARCELLINI, A., DAMINELLI, R., TENTO, A., FRANCESCHINA, G., and PAGANI, M. (2001), *The Umbria-Marche Microzonation Project: Outline of the Project and the Example of Fabriano Results*. Special issue: *The 1997-98 Umbria-Marche Earthquake*, Italian Geotechnical Journal. *XXXV*, (2), 22–29.

MASSIMINO, M. R., MAUGERI, M., and ZUCCARELLO, S. (2001), *The Grade-2 Microzonation of Sellano*. Special issue: *The 1997-98 Umbria-Marche Earthquake*, Italian Geotechnical Journal. *XXXV*, (4), 79–96.

MASSIMINO, M.R., and MAUGERI, M. (2001), *The grade-3 microzonation of Noto*. Proc. Symp. *Geotechnical Analysis of Seismic Vulnerability of Historical Monuments*, Catania (Italy), November 2001.

MAUGERI, M. and CARRUBBA, P. (1985), *Microzoning using SPT data*. Proc. X I.C.S.M.F.E., San Francisco, August 11–15, 1985. (*4*), 1831–1836.

MAUGERI, M., and FRENNA, S.M. (1995), *Soil response analysis for the 1990 southeast Sicily earthquake*. Third Internat. Conf. *Recent Advances in Geotechnical Earthquake Engineering and Soil Dynamics*, St. Louis, *2*, pp. 653–658.

MAUGERI, M. and CARRUBBA, P. (1997), *Microzonation for ground motion during the 1980 Irpinia earthquake at Calabritto, Italy*. XIV ICSMFE, Special Volume of ISSMFE TC4: *Earthquake Geotechnical Engineering*, Hamburg, Germany, 1997, September *6–12*, pp. 81–96.

MAUGERI, M. and CAVALLARO A. (2001): "Caratterizzazione Dinamica dei Terreni di Augusta e Noto"; *Scenari di Pericolosità Sismica ad Augusta, Siracusa e Noto*. A cura di L. Decanini e G. F. Panza. CNR-Gruppo Nazionale per la Difesa dai Terremoti, Roma, 65–79.

MEDVEDEV, J., Engineering Seismology (Academia Nauk Press, Moscow (1962)), 260 pp.

MEDVEDEV, S.V. (1960), *The forecast of seismic effects on constructions*. Proc. of Second World Conference on Earth. Engineering Japan.

MORA, S., and VAHRSON, W. (1994), *Macrozonation Methodology for Landslide Hazard Determination*, Bull. Assoc. Eng. Geologists XXXI, (1), 49–58.

REGLES, PS 92 (1992), *Règles de construction parasismique, Règles PS applicables aux bâtiments, dites Règles PS 92*. Norme française, AFNOR 1995, 217 pp.

SCANDONE, P., PATACCA, E., MELETTI, C., BELLATALLA, M., PERILLI, N., and SANTINI, U. (1992), *Struttura Geologica, evoluzione cinematica e schema sismotettonico della Penisola Italiana*. Atti del convegno GNDT, Pisa 25-27 Giugno 1990, *1*, 119-135. *Edizione Ambiente, Bologna*.

SHIMA, E. (1978), *Seismic Microzoning Map of Tokyo*, Proc. second Internat. Conf. of *Microzonation*. *1*, pp. 433–443.

TC4 (1993), *Manual for Zonation on Seismic Geotechnical Hazards.* Technical Committee for Earthquake Geotechnical Engineering, TC4, of I.S.S.M.F.E. Ed. The Japanese Society of Soil Mechanics and Foundation Engineering, December 1993, 149 pp.

TENTO, A., DE FRANCO, R., FRANCESCHINA, G., and PAGANI, M. (2001), *Site Effect Zonation of the Fabriano Municipality.* Special issue: *The 1997-98 Umbria-Marche earthquake,* Italian Geotechnical Journal, *XXXV*, (2), 136–151.

(Received July 16, 2002, revised August 12, 2003, accepted September 9, 2003)

To access this journal online:
http://www.birkhauser.ch

Pure appl. geophys. 162 (2005) 683–697
0033–4553/05/040683–15
DOI 10.1007/s00024-004-2632-0

❙ Pure and Applied Geophysics

On the Observed Intensity Filtering in the Anisotropic Distribution Modelling of Macroseismic Intensity

D. TERMINI[1], A. TERAMO[1], T. TUVÈ[1] and A. BOTTARI[1]

Abstract—The anisotropic modelling of intensity distribution, affected by the construction of macroseismic planes, allows an analysis of the influence of each point of observed intensity on the analytical determination of epicenter and of the principal attenuation directions. Such a procedure is a vital aid in the cases in which the observed intensity points, that, for location or joined intensity level, are not consistent with an anisotropic model of intensity attenuation. A suitable filtering on intensity levels associated with the points of the intensity map, for a better modelling of observed intensity distribution, is proposed with the aim of a better seismic hazard evaluation.

Key words: Macroseismicity, observed intensity filtering, macroseismic planes.

1. Introduction

The seismic intensity scales mainly used in Europe, such as the MCS scale (Mercalli-Cancani-Sieberg) allow the earthquake intensity in a given site to be evaluated by the comparison of the effects observed by a set of macroseismic indicators that characterizes degrees of scale of reference. The character widely qualitative/descriptive of the macroseismic indicators and their dishomogeneity does not allow the seismic intensity as a continuous quantity to be considered. Nevertheless, for decades it has been widely used to consider intermediate values of intensity (III-IV, IV-V, etc.) and in seismic hazard evaluation to individualize the expected intensity by half degrees. Many studies and catalogues of various periods assign noninteger values of intensity to single points and/or certain isoseismal areas (BARATTA, 1901; MEZCUA, 1982; POSTPISCHL, 1985; CECIC *et al.*, 1996; CAMASSI and STUCCHI, 1997). Such a practice is not always justified and it could determine ambiguity. In fact, half degrees are used both by authors (the majority) in cases in which the attribution of an integer intensity value is uncertain and by others in an attempt to refine the intensity determination. The last case happens when authors consider that the observed macroseismic effects are in some way intermediate between those that characterize two close degrees of intensity.

[1]Osservatorio Sismologico, Università di Messina, Via Osservatorio, 4 – 98121, Messina, Italy.
E-mail: dtermini@unime.it

The procedure proposed here consists of two different approaches towards a redefinition of observed intensity points.

The first one, addressed to the approximation of non-integer intensity levels to the immediately higher or lower integer ones, originates from the need for a macroseismic intensity distribution that presents only integer levels, for a better evaluation of intensity attenuation and, especially, in seismic hazard analyses, for an easier comparison between the observed intensity and the expected one. It is observed, moreover, that the non-integer intensity levels in an intensity distribution modelling, such as the macroseismic plane (TERAMO *et al.*, 1996), introduce an error level that is reflected on the determination of the epicenter and the principal attenuation directions. The macroseismic planes organized only by integer intensity levels will be called *1st-level filtered macroseismic planes*.

The second approach is oriented to a reconfiguration of integer intensity levels by a spatial distribution analysis of intensity levels, verifying their consistency with an

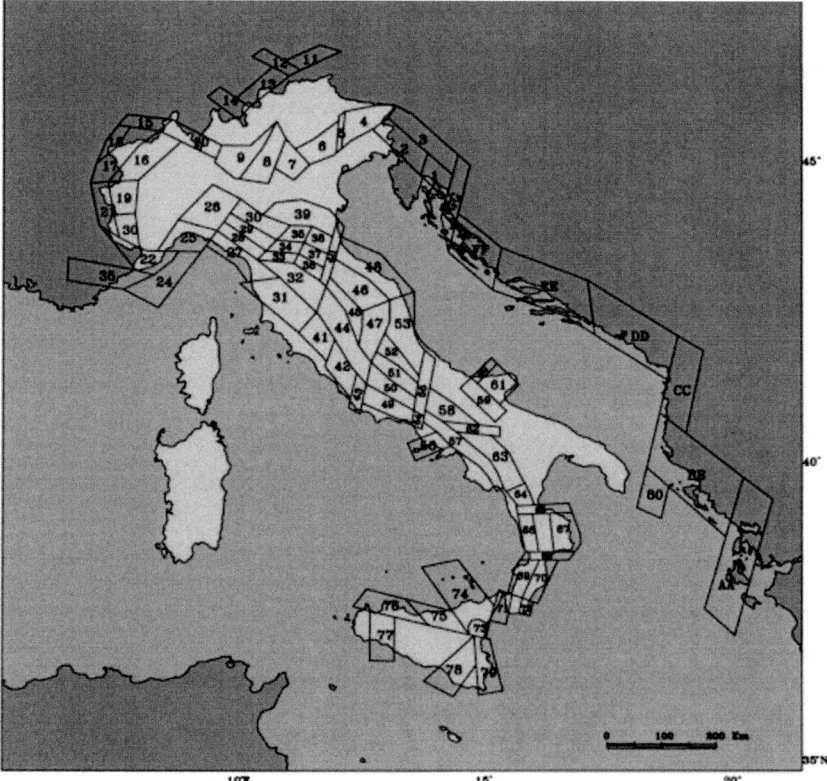

Figure 1
Seismogenetic zonation (SCANDONE *et al.*, 1992), adopted by G.N.D.T. for the seismic hazard evaluation of the Italian region.

anisotropic model of intensity attenuation. The aim is to model, for a given seismic event, a distribution of intensity levels to be assumed as a reference for the determination of attenuation coefficients by an anisotropic law (TERAMO *et al.*, 1995) and, for all the events of a seismogenic zone, a better definition of the virtual intensity that is the distribution of reference for seismic hazard evaluation (TERAMO *et al.*, 1998, 1999). The *1ˢᵗ-level filtered macroseismic planes*, reconfigured by a suitable filter on integer intensity levels, will be called *filtered macroseismic planes*.

The procedure that is proposed has been tested on several seismic events belonging to different seismogenic zones individualized by the numbers 55, 56, 58, 63, 71 and 78 in the seismogenic zonation of Italian territory reported, in the last version of April 1996 (SCANDONE *et al.*, 1992) (Fig. 1).

2. The Approximation of the Half Intensity Levels

2.1 The Usual Procedures

The approximation of half intensity levels of an intensity map of a seismic event to the integer ones can be carried out using the usual procedures that, respectively:

1. consider the half intensity levels as independent ones;
2. exclude the half intensity levels;
3. approximate all the half intensity levels to the immediately higher integer ones;
4. approximate all the half intensity levels to the immediately lower integer ones.

The usual approximation procedures of the half intensity levels to the integer ones have been applied to the intensity maps of the earthquakes belonging to the seismogenic zones indicated and the corresponding macroseismic planes have been designed. For reasons of brevity, only those relative to the Campobasso earthquake on 26-07-1805 (Fig. 2) of seismogenic zone N. 58 are shown (Figs. 3–6).

The reliability of these approximation procedures and, consequently, of each macroseismic plane has been evaluated testing the consistency of the epicenter and the principal attenuation directions, determined by a suitable vectorial modelling (TERAMO *et al.*, 1993, 1995a, 1996; BOTTARI *et al.*, 1995; TERMINI *et al.*, 1995) that involve all intensity levels, respectively with the barycenter of the areolas of the macroseismic planes to which the maximum intensity level is associated and the tectonic trend of the seismogenic zones to which the events are referable.

The results obtained for the events listed in Table 1 are reported in Table 2, in which a comparison is highlighted between the E epicenters and the *t* minimum attenuation directions (calculated with the 1, 2, 3, 4 described approximation procedures of half intensity levels), respectively with the *E′* barycenters of the areolas of the macroseismic planes to which the maximum intensity level is associated and *t′* directions representative of the tectonic trend of each seismogenic zone.

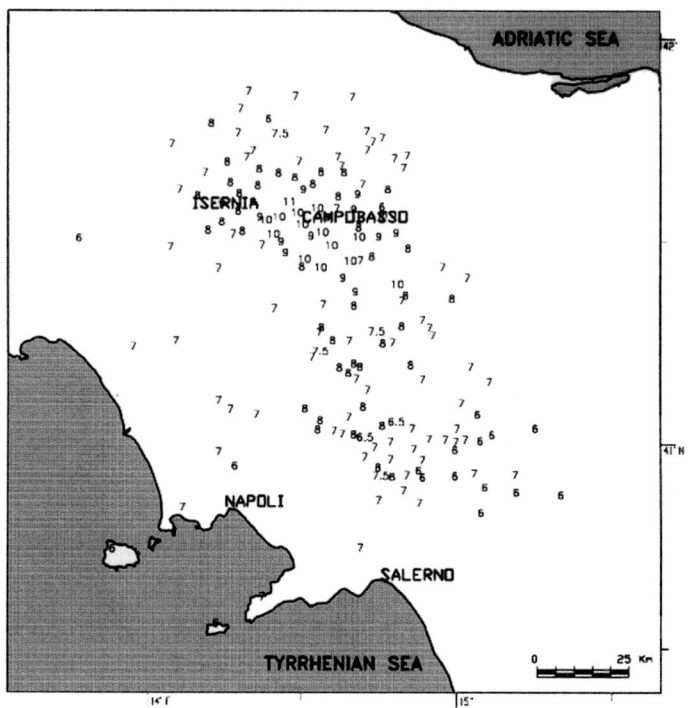

Figure 2
Intensity map of 26-07-1805 Campobasso earthquake (ESPOSITO *et al.*, 1987).

In general, it is observed that the standard deviations of E epicenters and t directions are generally greater than the E' barycenters and t' directions, owing to the greater number of approximations of half intensity levels that involve all intensity levels of a macroseismic plane.

This means that, by these approximations of half intensity levels, an intensity distribution is characterized that is not consistent with the one that an anisotropic model of intensity attenuation (TERAMO *et al.*, 1995b) would allow to be prefigured.

Therefore, it is necessary to shape a different approximation procedure of the half intensity levels, considering especially that these uncertainties are never of the same sign. In fact, it is hardly credible that, in the construction of intensity maps, the damage has been always under- or overestimated.

Here a new approximation procedure of half intensity levels is proposed which, by geometric considerations, allows the differentiation of approximation of each intensity level to the immediately higher or lower integer one.

2.2. *The Proposed Procedure*

The procedure that is proposed represents the first step of a filtering of the observed intensity levels. It is to be considered that the approximation of half

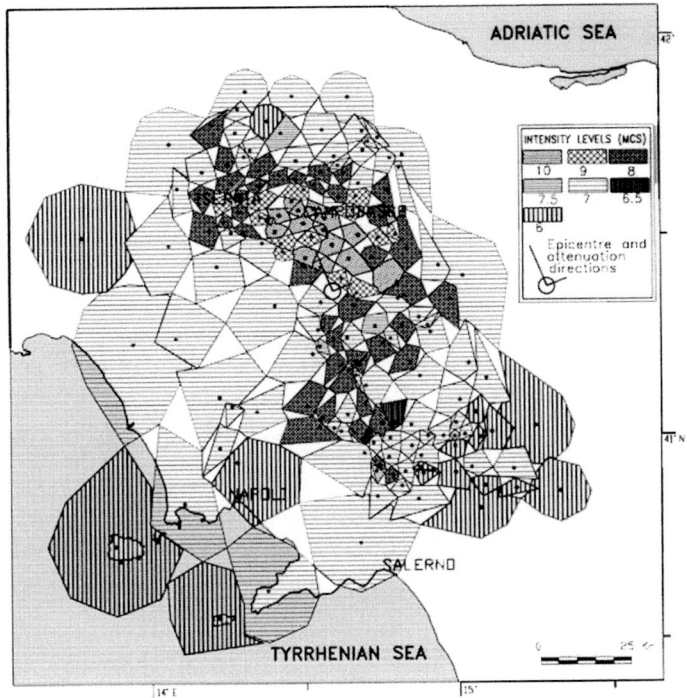

Figure 3
26.07.1805 Campobasso earthquake: Epicenter and principal attenuation directions calculated through the macroseismic plane designed considering half degrees as distinct intensity levels.

intensity levels to the immediately higher or lower integer ones, is restricted to 0.5 MCS degrees and does not undermine in any way eventual corrections to the observed intensity points relative to the local seismological conditions that involve differences equal to 1,5 or 2 MCS degrees.

The proposed procedure begins with the consideration of the probability that a point, called P_m, representative of a non-integer observed intensity level, is approximate to the immediately higher integer intensity level of a P_s point or to the lower one of a P_i point, is as greater or as smaller as the distance of P_m from P_s or from P_i.

Therefore, for an intensity map of an event, all the half intensity levels P_m points are considered and their distances from the nearest surrounding points representative of the immediately higher or lower integer intensity levels are calculated. The generic P_m point will be approximated to the immediately higher or lower integer intensity level according to the shortest distance from the point representative of the immediately higher or lower integer intensity level. This procedure is explained in Figure 7 where it is possible to verify that the half intensity level 7.5 MCS P_m point is approximated to the immediately higher integer one 8.0 MCS.

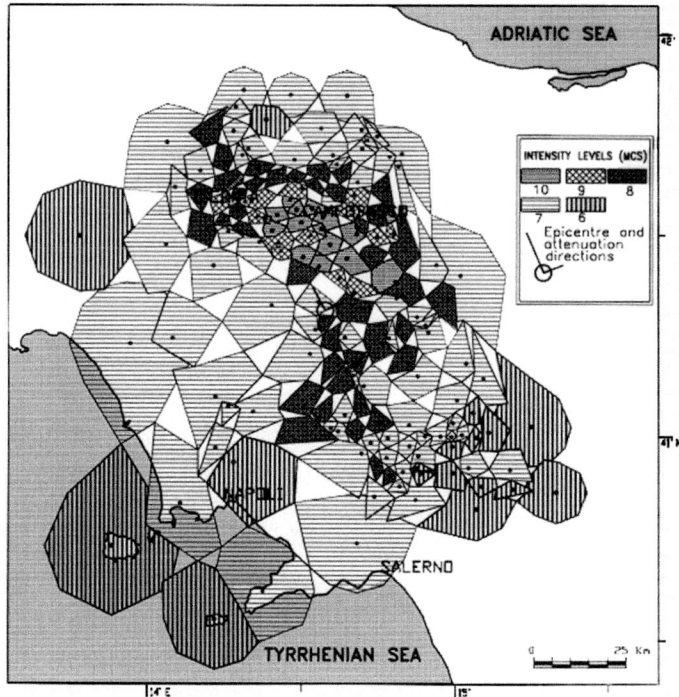

Figure 4
26.07.1805 Campobasso earthquake: Epicenter and principal attenuation directions calculated through the macroseismic plane designed excluding the half degrees.

The application of this procedure to the intensity map allows the design of a macroseismic plane constituted only by integer intensity levels that will be called 1st–*level filtered macroseismic plane*. Figure 8 shows the 1st–level filtered macroseismic plane of the Campobasso earthquake (26-07-1805).

The reliability of this new procedure has been evaluated verifying (Table 3), for the events listed in Table 1, the consistency of the E epicenters and the t directions of minimum attenuation, determined with the above mentioned vectorial procedure that involves all intensity levels, respectively with the E' barycenters of the areolas of the macroseismic planes to which the maximum intensity level is associated and the t' directions representative of the tectonic trend of each seismogenic zone.

The use of the proposed procedure points out that, compared to the original one, the obtained intensity distribution is more consistent with the one that an anisotropic model of intensity attenuation would allow to be prefigured. In particular, it is to be observed that the determination of epicenters and the attenuation directions of each intensity distribution represent a double level of verification of the proposed procedure with different characteristics. In fact, for the first one, greater differences are acceptable because the points of observed intensity might not define correctly the

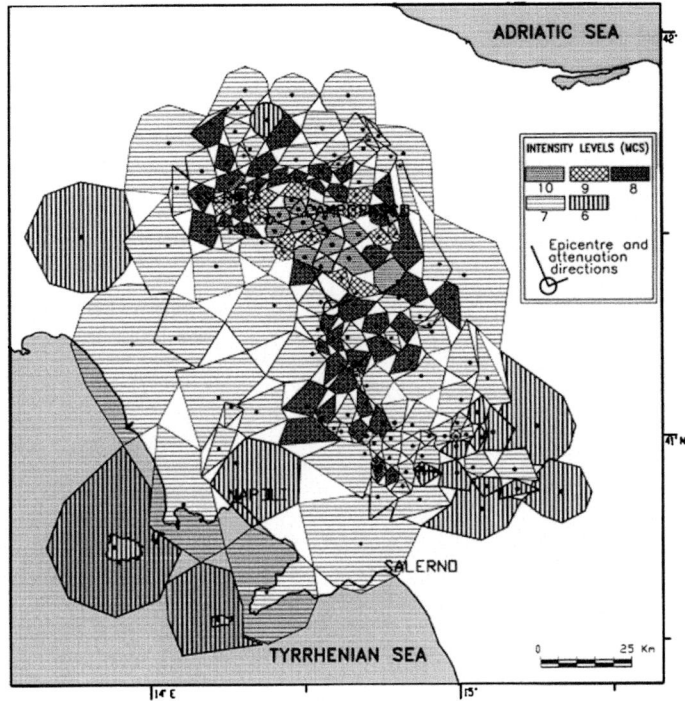

Figure 5
26.07.1805 Campobasso earthquake: Epicenter and principal attenuation directions calculated through the macroseismic plane designed approximating the half degrees to higher integer intensity level.

barycenter of the area to which the maximum intensity level is associated. For the second one, on the contrary, only slight differences relative to the minimum attenuation directions of macroseismic intensity are acceptable, considering that this is a considerably more significant result because it is strictly connected with the geological model of the analyzed seismogenic zone.

In any case, it is observed that the particular shape of the corresponding macroseismic plane needs a further reshaping aimed at a redistribution of the integer levels of intensity, consistent with a spatial distribution of intensity levels, starting from the epicenter and with reference to the geological model of the seismogenic zone. All this is to obtain an intensity distribution that can be considered as the reference one to be used in seismic hazard studies.

3. The Reconfiguration of Integer Levels of Observed Intensity

3.1. The Proposed Filtering

The considerations with which we started for the characterization of the filter, through which to carry out an analysis of the spatial distribution of integer intensity

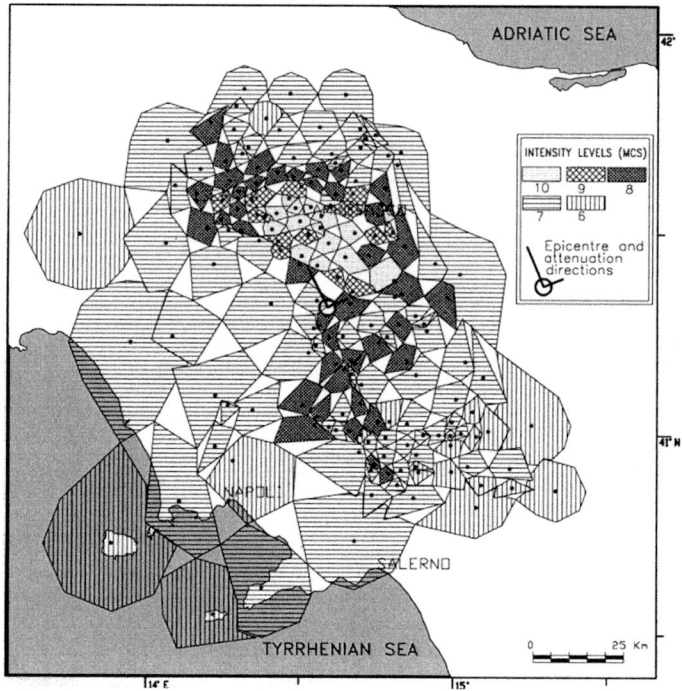

Figure 6

26.07.1805 Campobasso earthquake: Epicenter and principal attenuation directions calculated through the macroseismic plane designed approximating the half degrees to the lower integer intensity level.

levels, are strictly connected with the need to have an easily usable instrument tool, based on easily verifiable geometrical elements.

The proposed filter, applied to the 1st-level filtered macroseismic planes, is configured correlating geometry of areolas to the corresponding distances from the

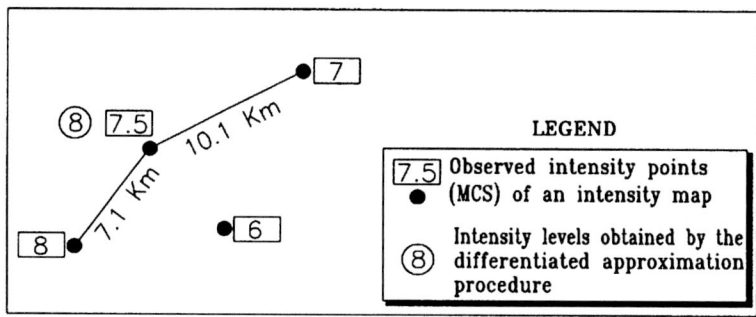

Figure 7

Approximation of half degrees procedure.

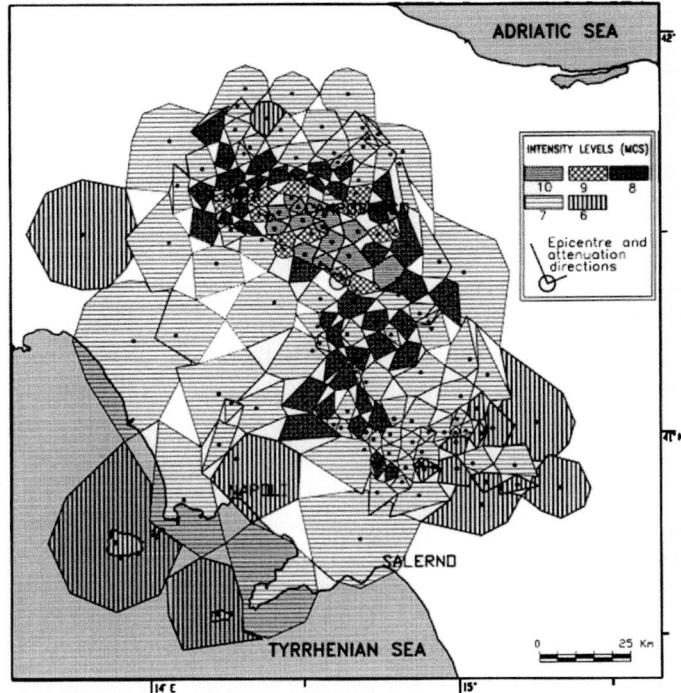

Figure 8

26.07.1805 Campobasso earthquake: Epicenter and principal attenuation direction calculated through the 1st –level filtered macroseismic plane.

epicenter and to associated intensity levels, within an anisotropic modelling of the macroseismic intensity attenuation.

It starts therefore with a reconfiguration of the macroseismic plane, obtained by moving from the areolas to which the lowest intensity levels are associated. In particular, it is to be prefigured that to a given areola built around the P_{ij} i-th point of the intensity map to which the j-th level of observed intensity is associated, a j^* intensity level, different from j, has to be associated, if :

1. the areola in study is surrounded, at least in 50% of its perimeter, by other areolas built around the $P_{i^*j^*}$ to which the j^*-th level of observed intensity is associated;
2. for $j^* < j$, denoting E - P_{ij} and E - $P_{i^*j^*}$ the corresponding distances from the epicenter of the P_{ij} and $P_{i^*j^*}$ points, we have E - $P_{ij} > E$ - $P_{i^*j^*}$;
The macroseismic plane, obtained renaming in such a way the areolas, will be named 2nd-level filtered macroseismic plane.

The consistency of such a further filter can be tested both by the usual double levels of verification, referable to the determination of the epicenter and principal directions of attenuation, and by the characterization of a *local virtual distribution* of the macroseismic intensity relative to an earthquake.

On this last subject, it is to be observed that the *regional* virtual distribution of intensity, originally proposed as a modelling of reference for all the events referable to the same seismogenic zone (TERAMO *et al.*, 1998), individualizes an intensity distribution that is characteristic of the seismogenic zone for seismic hazard evaluations.

By the same procedure, it is also possible to configure a *local* virtual intensity distribution that differs from the *regional* one because it concerns only an earthquake, which gives a distribution of the intensity levels consistent with an anisotropic modelling of macroseismic intensity attenuation (TERAMO *et al.*, 1999).

On the analogy of the *regional one*, the *local* virtual intensity distribution is characterized by a virtual epicenter and virtual directions of maximum and minimum attenuation. The first one is individualized by the C center of v_i vectors system of maximum and minimum attenuation (TERAMO *et al.*, 1996) deduced from the *2nd– level filtered macroseismic plane* of the earthquake in study, the virtual attenuation directions coincide with the central axes of the v_i vectors systems. Geometrically it is represented by a set of similar ellipses, to each of which an intensity level is associated, whose centers coincide with the C virtual epicenter and whose semiaxes, oriented along virtual directions of maximum and minimum attenuation, are proportional to the ρ_0 semiaxes of the first ellipsis, by K^i amplification factors (TERAMO *et al.*, 1998).

For a given seismic event the superimposition of local virtual intensity distribution to the 2nd level filtered macroseismic plane represents a significant test of the consistency of the adopted filter, because it allows eventual overestimates or underestimates of observed intensity to be highlighted, although it characterizes itself as an efficient instrument for individualizing eventual anomalies of propagation directly referable to particular local geomorphological conditions, that can be detected by a careful survey *in situ*.

It is to be noticed moreover, that the adoption of such a modelling contributes to a better definition of the regional virtual intensity distribution of a given seismogenic zone that, being carried out by 2nd level filtered macroseismic planes of the earthquakes, will determine more accurate calculations both of virtual epicenter and virtual directions of maximum and minimum attenuation, and regional attenuation coefficients, carried out by an anisotropic attenuation law.

Table 1

List of the analyzsed earthquakes grouped for seismogenic zones

Earthquake	Date (dd.mm.yy)	Intensity (MCS)	Reference	Seismogenic Zone
Lucania	08-09-1694	X	POSTPISCHL, D. (1985)	63
Irpinia	14-03-1702	X	POSTPISCHL, D. (1985)	56
Campobasso	26-07-1805	XI	ESPOSITO *et al.* (1987)	58
Lama dei Peligni	26-09-1933	IX	POSTPISCHL, D. (1985)	55
Alfedena	07-05-1984	VII -VIII	CARRARA *et al.* (1985)	55

3.2. Analyses and Results

The proposed modelling has been applied, furthermore, to the seismic events indicated in Table 1, to many earthquakes in southern Calabria, Eastern Sicily and, in a particular way, to the earthquake of the Messina Straits on 28-12-1908 (BOTTARI *et al.*, 1986). For brevity, only the results relative to the earthquake of 26-07-1805 in Campobasso are reported, whose *1st –level filtered macroseismic plane* (Fig. 8) has been reconfigured by the proposed filter, with reference to the integer intensity levels. The corresponding *2nd –level filtered macroseismic plane* (Fig. 9), compared with the 1st level ones, highlights a different spatial distribution of intensity levels.

Verifications concerning the reliability of the proposed modelling have been arranged using different analytical type procedures, moving from the geometry of the 2nd –level filtered macroseismic plane. The first two allow an analytical verification of consistency of the epicenter and principal attenuation directions calculated, respectively, with the area of maximum shaking and the structural lineaments of the seismogenic zone in study. The third derives from the superimposition of the two intensity distributions carried out, respectively, from the 2nd –level filtered

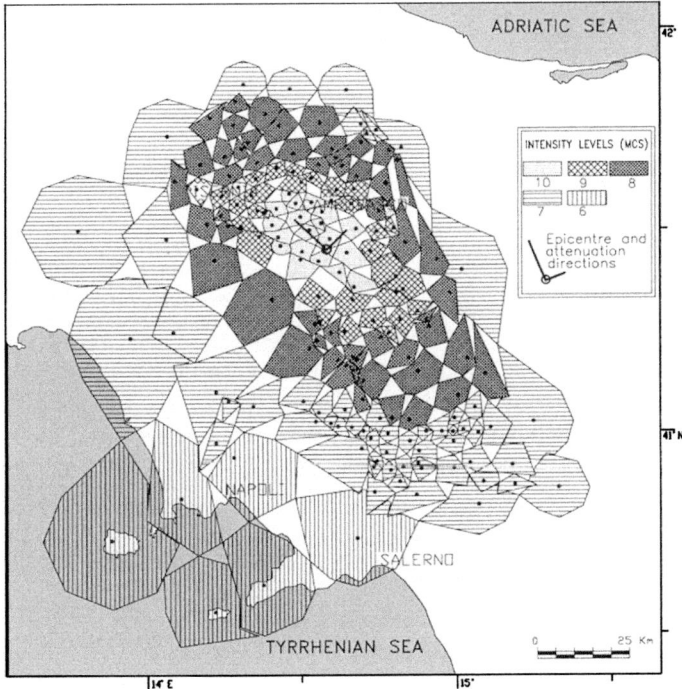

Figure 9

26.07.1805 Campobasso earthquake: Epicenter and principal attenuation direction calculated through the 2nd –level filtered macroseismic plane.

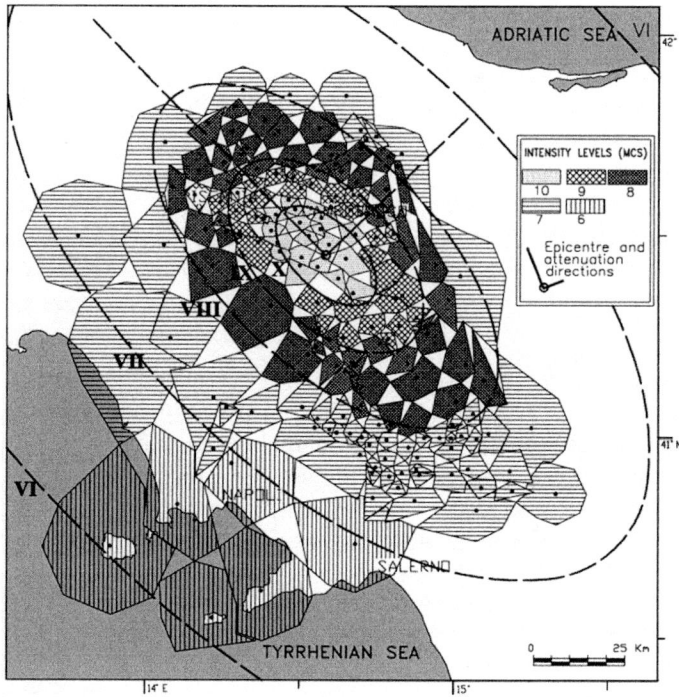

Figure 10
26.07.1805 Campobasso earthquake: Local virtual intensity distribution superimposed to 2nd –level filtered macroseismic plane.

macroseismic planes and the local virtual intensity distributions of the same earthquake (Fig. 10).

The results obtained, in comparison with the corresponding ones shown in Table 2, highlight epicentral coordinates (137.19; 291,62 ; x,y [km]) very close to the area of maximum shaking and directions of maximum and minimum attenuation (44 °—anticlockwise from the North) consistent with the ones of structural lineaments of the seismogenic zone of the earthquake in study (NW-SE). In particular, the superimposition of the local virtual intensity distribution to the 2nd - level filtered macroseismic plane shows a good consistency of each local virtual isoseismal line with the sets of areolas associated to the corresponding intensity levels.

4. Conclusions

In the modelling of intensity distribution of historical earthquakes, the use of the 1st and 2nd –level filtered macroseismic planes represents a significant instrument

Table 2

Comparison between the E epicenters (°) and the t minimum attenuation directions (°°) calculated with the usual approximation procedures of half intensity levels (°°°), respectively with the E' barycenters (°) of the areolas of the macroseismic planes to which the maximum intensity level is associated and t' directions representative of the tectonic trend of each seismogenetic zone, for some earthquakes of Seismogenetic Zones of Central - southern Italy.

S. Z.	Earthquake	E x, y ; (km) by 1/ 2/3/4	E' x, y ; (km) by 1/2/3/4	t by 1/2/3/4	t'
55	Lama dei Peligni (1933)	89.76, 341.22	106.15, 338.42	13.89 N	NNW-SSE
		87.93, 344.11	104.65, 337.36	37.00 N	
		91.33, 339.76	106.15, 338.42	39.62 N	
		88.64, 342.79	106.15, 338.42	36.95 N	
55	Alfedena (1984)	88.98, 296.15	101.11, 298.99	23.83 N	NNW-SSE
		84.36, 289.13	89.11, 297.62	28.38 N	
		87.94, 289.26	100.94, 298.40	22.66 N	
		87.88, 288.06	89.24, 299.59	25.90 N	
56	Irpinia (1702)	177.79, 225.29	169.15, 233.85	71.96 N	WNW-ESE
		161.79, 234.18	168.54, 233.78	77.27 N	
		175.08, 228.45	166.51, 234.14	75.01 N	
		159.23, 232.19	169.02, 233.73	64.01 N	
58	Campobasso (1805)	138.18, 260.71	137.99, 275.78	22.92 N	NW-SE
		127.84, 293.79	129.33, 267.97	28.04 N	
		137.53, 257.62	137.99, 275.78	26.81 N	
		137.53, 257.50	137.99, 275.78	27.12 N	
63	Lucania (1694)	203.80, 213.91	215.40, 201.85	355.58 N	NNW-SSE/ NW-SE
		182.62, 211.26	216.03, 215.47	14.35 N	
		186.73, 207.34	215.03, 201.89	20.19 N	
		183.91, 211.13	215.73, 216.83	11.73 N	

(°) *O,,x,y reference system: Origin: O ≡ (39°N LAT, 13°E LONG) ; xaxis: (O, East); yaxis: (O, North);*
(°°) *Anticlockwise from North*
(°°°) *Usual approximation procedures that, respectively:*
1—*consider the half intensity levels as independent ones*
2—*exclude the half intensity levels*
3—*approximate the half intensity levels to the immediately higher integer ones*
4—*approximate the half intensity levels to the immediately lower integer ones*

within the set of analyses aimed at seismic hazard evaluation. The adoption of filters concerning intensity levels allows in fact, by the same structure of the macroseismic plane, specific evaluations on each point of observed intensity, highlighting eventual intensity amplification, referable to particular local geological conditions, or evaluation errors. All the analyses carried out on a large number of earthquakes belonging to different seismogenic zones show a better evaluation of the spatial distribution of intensity levels, consistent with an anisotropic modelling, and in consequence, a more correct analytical determination both of the epicentral coordinates and the directions of maximum and minimum attenuation of intensity.

It is to be noticed moreover that, for strong earthquakes, the large number of observed intensity points makes the effect of filters more transparent and the consistency verification clearer. For weak earthquakes, the configuration even of a low number of observed intensity points, highlights the efficacy of the proposed procedure.

The characterization of the local virtual intensity distribution of every seismic event represents, moreover, an important reference for a more correct verification of consistency between the observed intensity distribution, that in any case experiences remarkable error levels, and the one analytically deduced with reference to parameters deduced from a critical analysis of the spatial distribution of observed intensity levels. The local virtual intensity distribution characterizes in fact a geometrical representation of the intensity levels distribution consistent with the 2nd –level macroseismic plane, that is with the epicenter and intensity attenuation directions and the structural lineaments of the seismogenic zone to which the earthquake in the study belongs.

REFERENCES

BARATTA, M. (1901), *I terremoti d'Italia. Saggio di storia, geografia e bibliografia sismica italiana*, Torino.

BOTTARI, A., CARAPEZZA, E., CARAPEZZA, M., CARENI, P., CEFALI, F., LO GIUDICE, E., and PANDOLFO, C. (1986), *The 1908 Messina Strait Earthquake in the regional Geostructural Framework*, J. Geodyn. *5*, 275–302.

BOTTARI, A., STILLITANI, E., TERAMO, A., and TERMINI, D. (1995), *Le incertezze dei dati macrosismici nella modellazione anisotropa dell'attenuazione*, Boll. Acc. Gioenia, Catania (Italy), Sci. Nat. *28*, 349, 259–274.

CAMASSI, R. and STUCCHI, M. (1997), *Nt4.1 un catalogo parametrico di terremoti di area italiana al di sopra della soglia del danno*, GNDT, Internal Report, 84pp.

CARRARA, C., FREZZOTTI, M., MARGOTTINI, C., MOLIN, D., and NARCISI, B. (1985), *Seismotectonic investigations on the earthquakes of May 7 and 11 1984 in Latium-Abruzzo area (Central Italy)*, Poster. 3rd Internat. Symp. on the *Analysis of Seismicity and Seismic Risk*, Liblice .

CECIC, I., MUSSON, R.M.W., and STUCCHI, M. (1996), *Do seismologist agree upon epicenter determination from macroseismic data ? A survey of the ESC "Macroseismology" Working Group*, Ann. Geofis. *39* (5), 1013–1027.

ESPOSITO, E., LUONGO, G., MATURANO, A., and PORFIDO, S. (1987), *Il terremoto di S. Anna del 26 Luglio 1805*, Memorie della Società Geologica Italiana, *37*, 171–191.

MEZCUA, J. (1982), *Catalogo general de isosistas de la Peninsula Iberica*, I.G.N., Madrid, Publicacion 202, 322pp.

POSTPISCHL, D. (ed.) (1985), *Catalogo dei terremoti italiani dall'anno 1000 al 1980*, Consiglio Nazionale delle Ricerche. Progetto Finalizzato Geodinamica, Roma, 239pp.

POSTPISCHL, D. (ed.) (1985), *Atlas of Isoseismal Maps of Italian Earthquakes*, C.N.R. Quad. Ric. Scient. *114*, 2A.

SCANDONE, P., PATACCA, E., MELETTI, C., BELLATALLA, M., PERILLI, N., and SANTINI, U. (1992), Struttura geologica, evoluzione cinematica e schema sismotettonico della Penisola Italiana. Att del convegno GNDT, Pisa 25–27 Giugno 1990, *1*, 119–135. Edizione Ambiente, Bologna

TERAMO, A., STILLITANI, E., and BOTTARI, A. (1993), *Le direzioni principali di attenuazione dell'intensità in una caratterizzazione anisotropa dei campi macrosismici*, Boll. Acc. Gioenia, Catania (Italy), Sci. Nat. *26*, 342, 383–395.

TERAMO, A., STILLITANI, E., and BOTTARI, A. (1995a), *Anisotropic Characterization of the Macroseismic Fields,* Natural Hazards *11(3)*, 223–245.

TERAMO, A., STILLITANI E., and BOTTARI A. (1995b): On an Anisotropic Attenuation Law of the Macroseismic Intensity, Natural Hazards *11*, 203–221.

TERAMO, A., TERMINI, D., and STILLITANI, E. (1995c), Riconfigurazione dei piani macrosismici mediante un'analisi della distribuzione spatiale dei livelli di intensita'. 14° Conv. Ann. C.N.R. - G.N.G.T.S, Roma, 1, 205–210.

TERAMO, A., TERMINI, D., STILLITANI, E., and BOTTARI, A. (1996), *The Determination of the Epicentre by a Vectorial Modelling of Macroseismic Intensity Distribution,* Natural Hazards *13*, 101–117.

TERAMO, A., TERMINI, D., STILLITANI, E., and BOTTARI, A. (1998), *An Anisotropic Modelling for the Determination of the Regional Attenuation Coefficients,* Natural Hazard *17*, 17–30.

TERMINI, D., TERAMO, A., STILLITANI, E., and BOTTARI, A. (1995) R*iconfigurazione dei piani macrosismici da un'analisi delle incertezze dei dati di intensità,* Boll. Acc. Gioenia, Catania (Italy), Sci. Nat. *28*, 349, 291–302.

(Received February 21, 2001, accepted January 10, 2002)

To access this journal online:
http://www.birkhauser.ch

Pure appl. geophys. 162 (2005) 699–705
0033–4553/05/040699–7
DOI 10.1007/s00024-004-2633-2

| Pure and Applied Geophysics

On the Intensity Virtual Area Characterization in the Intensity Distribution Modelling by Macroseismic Planes

DOMENICA TERMINI[1], ANTONIO TERAMO[1], and CARLA BOTTARI[1]

Abstract — A procedure is proposed for the reconfiguration of the macroseismic planes relative to earthquakes that, being characterized by a reduced number of points of observed intensity due to a lack of information, or having the epicenter very close to the coastline, are characterized by an incomplete distribution of observed intensity levels. The design of a plurality of virtual areas, through which a distribution of intensity consistent with an anisotropic model of attenuation is depicted, allows a reliable determination of macroseismic parameters of the same seismic event.

Key words: Macroseismic planes, intensity virtual areas.

1. Introduction

The proposed procedure produces a reconfiguration of the area distribution of the observed intensity levels of a given earthquake previously modelled by a macroseismic plane (TERAMO *et al.*, 1996). Such a reconfiguration, relative to earthquakes which have occurred in areas that have a remarkable lack of observed data, owing to the particular morphology of the territory, or because of the closeness of the epicenter to the coastline, is characterized by the design of virtual isoseismal lines that complete the intensity distribution individualized by the original intensity map of the considered event.

The need of such virtual areas is strictly correlated to the same modelling of macroseismic intensity distribution carried out through macroseismic planes, that can be considered reliable if the epicenter and the calculated attenuation directions (TERAMO *et al.*, 1995a; 1996) are consistent with the maximum shaking area of the earthquake in study and the structural lineaments of the area in study, respectively, or if the macroseismic plane has been designed with reference to an area distribution of intensity levels so as to ensure the invariance of epicenter and attenuation directions.

The procedure has been tested on several seismic events that took place in the seismogenic zones singled out from numbers 71, 72, 74, 75, 78, 79 in the seismic

---
[1] Osservatorio Sismologico, Università di Messina, Via Osservatorio, 4 – 98121Messina, Italy.
Email: dtermini@unime.it

zonation of Italian territory adopted from GNDT (SCANDONE *et al.*, 1992), but for the sake of brevity, only the results relative to the Messina earthquake on 28-12-1908 located in zone 71 which, even if it has a high number of observed intensity points, is characterized by their incomplete distribution referable to the location of the epicenter that is very close to the coastline.

2. Characterization of Intensity Virtual Areas

The modelling of the distribution of intensity levels of the 28-12-1908 earthquake, used for the design of virtual areas, is the 2nd-level filtered macroseismic plane (TERMINI *et al.*, 2005a) (Figs. 1,1a,1b) deduced from the original intensity map renaming the intensity levels associated with whole observed intensity points not consistent with their distance from the epicenter. Such a macroseismic plane characterizes an area distribution of intensity levels consistent with an anisotropic attenuation model (GRANDORI *et al.*, 1987; MAUGERI *et al.*, 1993; PERUZZA, 1995; TERAMO *et al.*, 1995b; TERMINI *et al.*, 2005b).

A verification of reliability of such a modelling for the analytical determination of the epicenter and principal attenuation directions (TERAMO *et al.*, 1995a; 1996) can easily be carried out, through a comparison of results which is obtained from specific calculations of those taken from literature (BARATTA, 1909, 1910; BOSCHI *et al.*, 1999; BOTTARI *et al.*, 1979,1986; CAMASSI and STUCCI 1997; SCHICK, 1977). In particular, it has been verified that both the epicenter and the attenuation directions are not

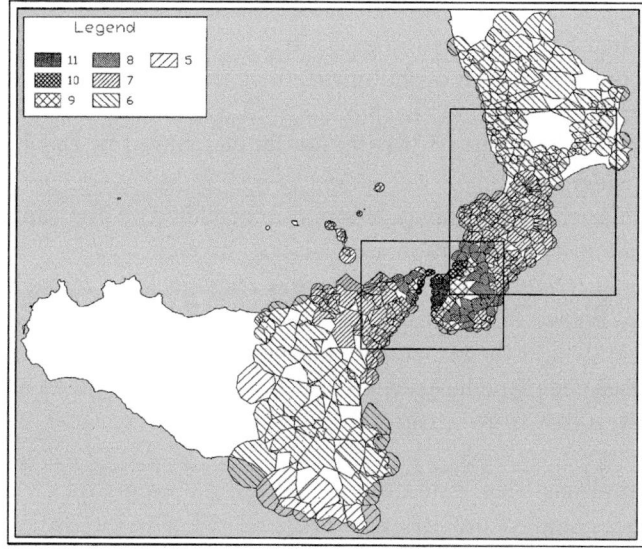

Figure 1
Messina earthquake on 28-12-1908: 2nd -level filtered macroseismic plane.

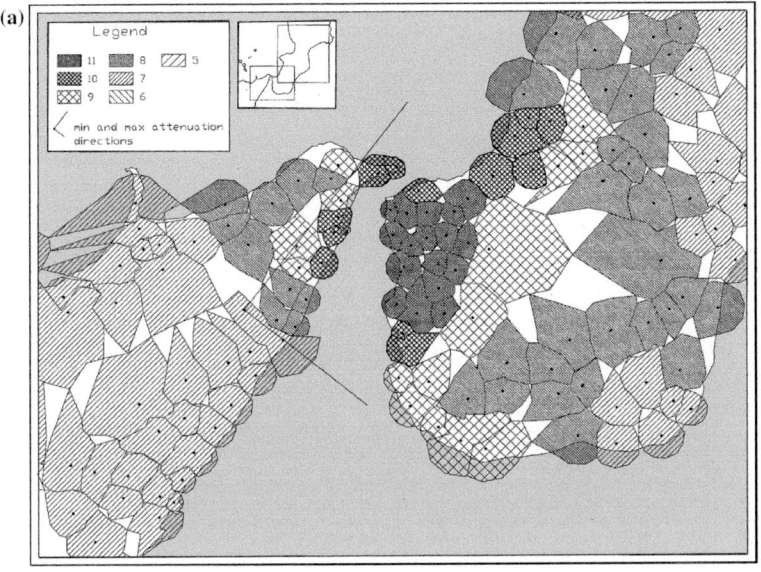

Figure 1a
Messina earthquake on 28-12-1908: 2nd -level filtered macroseismic plane (A macro).

consistent with the area of maximum shaking and the directions of structural lineaments of the seismogenic zone in study taken from the literature respectively.

From Figure 1a. it is possible to observe that the calculated epicenter is located on the northeastern Sicily coast, instead of the southwestern Calabrian one, about 30 km from the area of maximum shaking of the earthquake, whose barycenter actually individualizes the observed epicenter. The direction of the calculated minimum attenuation forms, on the contrary, an angle with the North direction equal to 322°N anticlockwise, which is not very consistent with that individualized by the structural lineaments: 330°N (BOTTARI et al., 1986).

The modelling of intensity distribution, carried out through the 2nd-level filtered macroseismic plane, reveals itself to be unsuitable for the determination of attenuation directions. It is necessary to turn to a redefinition of the macroseismic plane through the characterization of intensity virtual areas that supplement the observed intensity distribution, close to the areas lacking in observed intensity data, coinciding, as in this case, with the sea zones surrounding the area of earthquake shaking. The design of such intensity virtual areas is realized with reference to the lines that envelope the homonymous points of observed intensity, depicted in accordance with ellipses that have centers coinciding with the observed epicenter of the earthquake. The entire modelling of intensity distribution is depicted through a plurality of isoseismal areas bounded, on one hand, by the ellipses relative to two subsequent levels of observed intensity, and on the other, by the areolas of filtered macroseismic plane relative to the same intensity levels (Fig. 2).

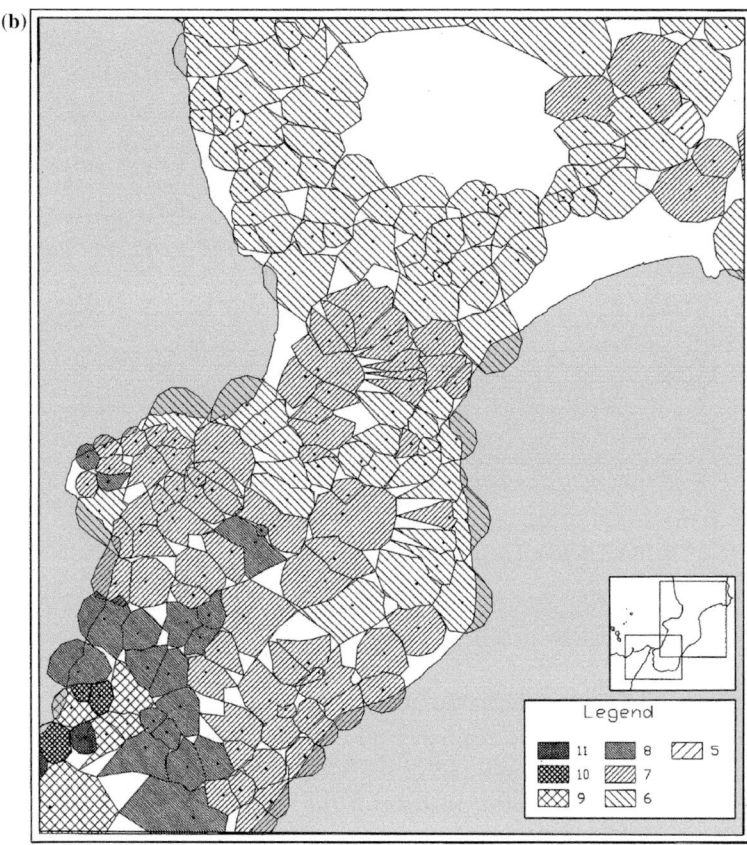

Figure 1b
Messina earthquake on 28-12-1908: 2nd -level filtered macroseismic plane (B macro).

From a comparison between Figures 1, 1a, 1b and 2 it is possible to verify that the entire distribution of intensity levels, carried out through the use of the intensity virtual areas and the areolas of the 2nd-level filtered macroseismic plane, is consistent with an anisotropic attenuation model of intensity considering, moreover, that the virtual isoseismal areas individualize a surface greater than that referable to the areolas of the corresponding macroseismic plane.

From a strictly analytical point of view, the efficacy of such a modelling is verifiable through the analytical determination of the epicenter and attenuation directions (Teramo *et al.*, 1996). The calculations carried out highlight in fact an epicenter correctly located close to the maximum shaking area (on the Calabrian shore of the Messina Straits) and the minimum attenuation directions (332°N anticlockwise) in good accordance with the directions of structural lineaments (330°N anticlockwise) of the Straits of Messina area (Fig. 3).

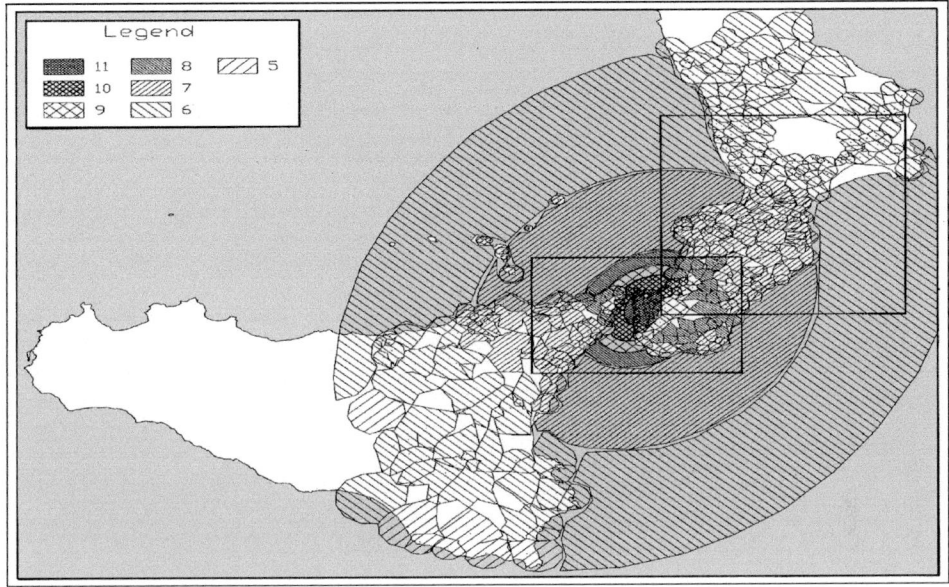

Figure 2
Messina earthquake on 28-12-1908: Observed intensity modelled through a 2nd -level filtered macroseismic
plane supplemented with virtual intensity areas.

A further confirmation of the reliability of the proposed modelling derives from the local virtual distribution of intensity of the earthquake (TERMINI et al., 2005b), characterized by a set of virtual ellipses that have the same center coinciding with the epicenter of the seismic event considered and semi-axes coinciding with the principal attenuation directions. Through a superimposition of such virtual ellipses of intensity to the intensity distribution of the earthquake, obtained supplementing the observed intensity, modelled through the 2nd-level filtered macroseismic plane with intensity virtual areas (Fig. 3), it is possible to verify the efficacy and the reliability of the proposed modelling, considering that the determination of an attenuation regional coefficient (TERAMO et al., 1998) necessary for the characterization of the seismic hazard of the region in study is based on such procedures.

3. Conclusions

The proposed procedure allows the determination of the epicenter and principal attenuation directions of the earthquakes which occurred in areas characterized by a lack of observation data, owing to both an unfavorable morphology of soils that over the centuries did not allow the spread of inhabited places, and the closeness of the coastline of the maximum shaking area of earthquakes.

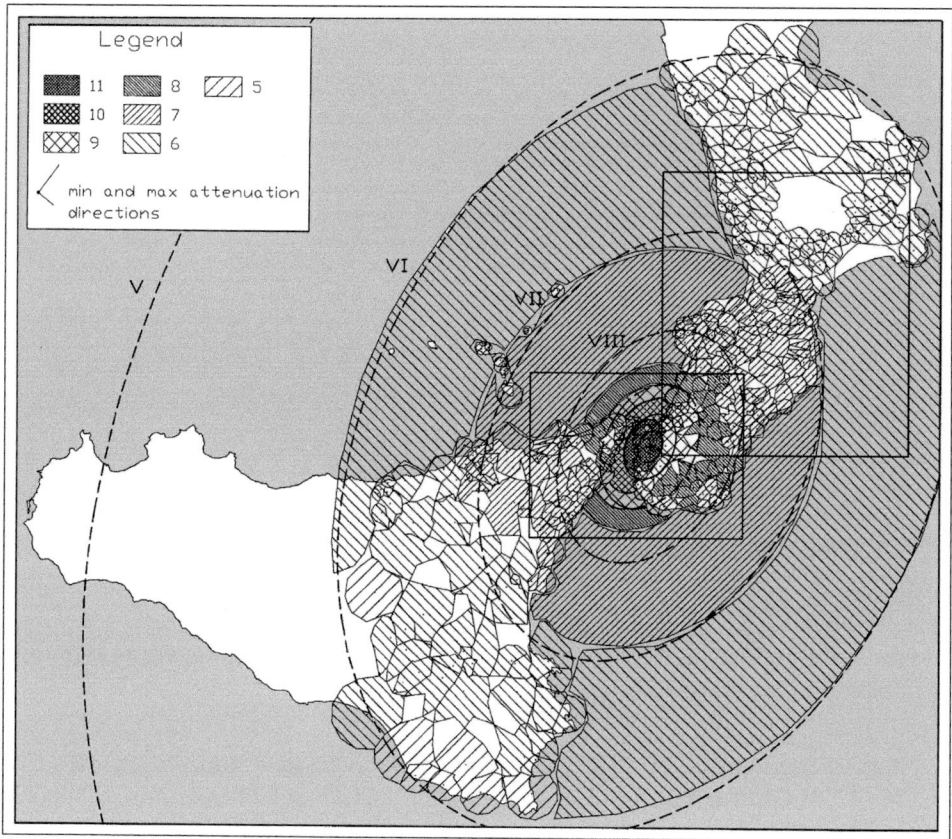

Figure 3
Local virtual intensity distribution for the Messina earthquake on 28-12-1908, superimposed to the 2nd-level filtered macroseismic plane supplemented with virtual intensity areas.

The characterization of virtual intensity areas that supplement the observed intensity, modelled through the 2nd-level filtered macroseismic plane, allows a suitable and more complete configuration of intensity level distribution. The consequent analytical determination of the epicenter and attenuation directions gives, in fact, results consistent with the maximum shaking area of the earthquake and the direction of structural lineaments respectively, allowing a specific reliability verification of the tracing of the isoseismal lines whose arbitrariness level would be hard to evaluate.

A further reliability verification of the use of virtual intensity areas derives from the verified consistency between the local virtual intensity distribution of earthquakes, and the distribution of intensity of the same earthquakes carried out supplementing the observed intensity, modelled through the 2nd-level filtered macroseismic plane, with intensity virtual areas.

REFERENCES

BARATTA, M. (1909), *Il terremoto calabro-siculo del 28 Dicembre 1908*, Boll. Soc. Geogr. It., S. IV, X, (*8*), 852–882; (*9*), 990–1010.

BARATTA, M. (1910), *La catastrofe sismica calabro-messinese (28 dicembre 1908)*, Relazione alla Soc. Geogr. It., Roma.

BOSCHI, E., GUIDOBONI, E., FERRARI, G., VALENSISE, G., and GASPERINI, P. (1999), *Catalogo dei forti terremoti in Italia dal 461 a.C. al 1990*, I.N.G., S.G.A. Storia Geofisica ambiente, **2**.

BOTTARI, A., FEDERICO, B., and LO GIUDICE, E. (1979), *Methodological Consideration Regarding the Determination of Some Macroseismic Field Parameters. Application to Earthquakes in the Calabro-Peloritan Arc*, Boll. Geof. Teor. Appl. *XXI* (*83*), 197–225.

BOTTARI, A., CARAPEZZA, E., CARAPEZZA, M., CARENI, P., CEFALI, F., LO GIUDICE, E., and PANDOLFO, C. (1986), *The 1908 Messina Strait Earthquake in the Geostructural Framework*, J. Geodyn. *5*, 275–302.

CAMASSI, R., and STUCCHI, M. (editors) (1997), *NT4.1 un catalogo parametrico di terremoti di area italiana al di sopra della soglia di danno*, GNDT, Internal Report, 84 pp.

GRANDORI, G., PERROTTI, F., and TAGLIANI, A. (1987), *On the attenuation of macroseismic intensity with epicentre distance*, Ground Motion and Engineering Seismology,(A.S. Camkak ed.),pp. 581–594, 3rd Int. Conf. on *Soil Dynamics and Earthquake Eng.*, Princeton, USA, 1986, (Elsevier, Amsterdam, 1987).

MAUGERI, M., MOTTA, E., and ZENG, X. (1993), *Attenuation Laws of Seismic Intensity in the Regions of Sicily and Calabria*, Soil Dynam. Earthq. Eng. *12*, 25–35.

PERUZZA, L. (1995), *Macroseismic intensity versus distance: constraints to the attenuation model*. In *Soil Dynamics and Earthquake Engeneering VII*, (Camkak A.S., Brebbia C.A. eds.) Comp. Mech. Publ., Southampton, 215–222.

SCANDONE, P., PATACCA, E., MELETTI, C., BELLATALLA, M., PERILLI, N., and SANTINI, U. (1992), *Struttura Geologica, evoluzione cinematica e schema sismotettonico della Penisola Italiana*. Atti del convegno GNDT, Pisa 25–27 Giugno 1990, *1*, 119–135. Edizione Ambiente, Bologna.

SCHICK, R.(1977), *Eine seismotektonische Bearbeitung des Erdenbens von Messina im Jahre 1908*. Geol. Jahrb., E *11*, 3–74.

TERAMO, A., STILLITANI, E.,and BOTTARI, A. (1995a), *Anisotropic Characterization of Macroseismic Fields*, Natural Hazard *11*, 223–245.

TERAMO, A., STILLITANI, E., and BOTTARI, A. (1995b), *An Anisotropic Attenuation Law of Macroseismic Intensity*, Natural Hazard *11*, 203–221.

TERAMO, A., TERMINI, D., STILLITANI, E., and BOTTARI, A. (1996), *The Determination of Epicentre by a Vectorial Modelling of Macroseismic Intensity Distribution*, Natural Hazard *13* ,101–117.

TERAMO, A., TERMINI, D., STILLITANI, E., and BOTTARI, A. (1998), *An Anisotropic Modelling for the Determination of Regional Attenuation Coefficients*, Natural Hazard *17*, 17–30.

TERMINI, D., BOTTARI, A., TERAMO, A., and TUVE', T. (2005a), *On the Observed Intensity Filtering in the Anisotropic Distribution Modelling of Macroseismic Intensity*, Pure Appl. Geophys. *162*, 683–697.

TERMINI, D., TERAMO, A., BOTTARI, C., and TUVÈ, T. (2005b), *An Anisotropic Attenuation Law of Macroseismic Intensity Performed on Virtual Intensity Distribution of Seismogenic Zones*, Pure Appl. Geophys., *162*, 707–714.

(Received July 7, 2001, accepted December 28, 2001)

 To access this journal online:
http://www.birkhauser.ch

Pure appl. geophys. 162 (2005) 707–714
0033–4553/05/040707–8
DOI 10.1007/s00024-004-2634-y

© Birkhäuser Verlag, Basel, 2005

▌Pure and Applied Geophysics

An Anisotropic Attenuation Law of Macroseismic Intensity Performed on Virtual Intensity Distribution of Seismogenic Zones

Domenica Termini[1], Antonio Teramo[1], Carla Bottari[1], and Tiziana Tuvè[1]

Abstract—An anisotropic attenuation law, based on an anisotropic characterization of intensity distribution for seismogenic zones, is proposed. This approach, that distinguishes itself for its consistency to the observed data, initially reconfigured by filtering procedures, is particularly suitable for seismic hazard evaluation.

Key words: Attenuation law, virtual intensity distribution, seismic hazard.

1. Introduction

A correct seismic hazard evaluation of a given area, that derives from a set of seismogenic zones, is strictly connected, as is known, to a careful evaluation of parameters that individualize the distribution and the attenuation of macroseismic intensity for each seismogenic zone (Grandori *et al.*, 1987; Maugeri *et al.*, 1993; Peruzza, 1995).

Studies carried out over the years (Teramo *et al.*, 1995a, b, c, d, 1996, 1998), based on a vectorial characterization of intensity distribution, allow a specific evaluation of observed data that, also through suitable filtering procedures (Termini *et al.*, 2005), both on the half and integer degrees, give a good configuration of macroseismic intensity distribution. The following determination of virtual intensity distribution (Teramo *et al.*, 1998), both *local* (relative to an earthquake) and *regional* (relative to all seismic events belonging to a given seismogenic zone), individualizes a distribution of reference that is characteristic both for the single earthquake, and for the set of earthquakes belonging to a seismogenic zone. Geometrically, such a distribution is represented by a set of similar ellipses, to each of which an intensity level is associated, with centers coinciding with the virtual epicenter of each seismic event or the seismogenic zone in study and whose semi-axes ρ_i, oriented towards the virtual directions of maximum and minimum attenuation, are proportional, by amplification factors K^i, to the semi-axes ρ_0 of the first ellipse.

[1] Osservatorio Sismologico, Università di Messina, Via Osservatorio, 4 – 98121 Messina, Italy.
E-mail: dtermini@unime.it

The analyses carried out on an area of located closeness between the Straits of Messina and eastern Sicily are relative to seismic events belonging to the seismogenic zones individualized by the numbers 71 and 79 in seismogenic zonation (SCANDONE *et al.*, 1992) adopted from GNDT (Gruppo Nazionale per la Difesa dai Terremoti, C.N.R., Rome) for the seismic hazard evaluation of Italy, in the last version of April 1996.

The characterization of an attenuation law performed on virtual intensity distribution of an earthquake or a set of earthquakes which originated from a given seismogenic zone, represents an effective instrument for seismic hazard evaluation and configuration of risk scenarios.

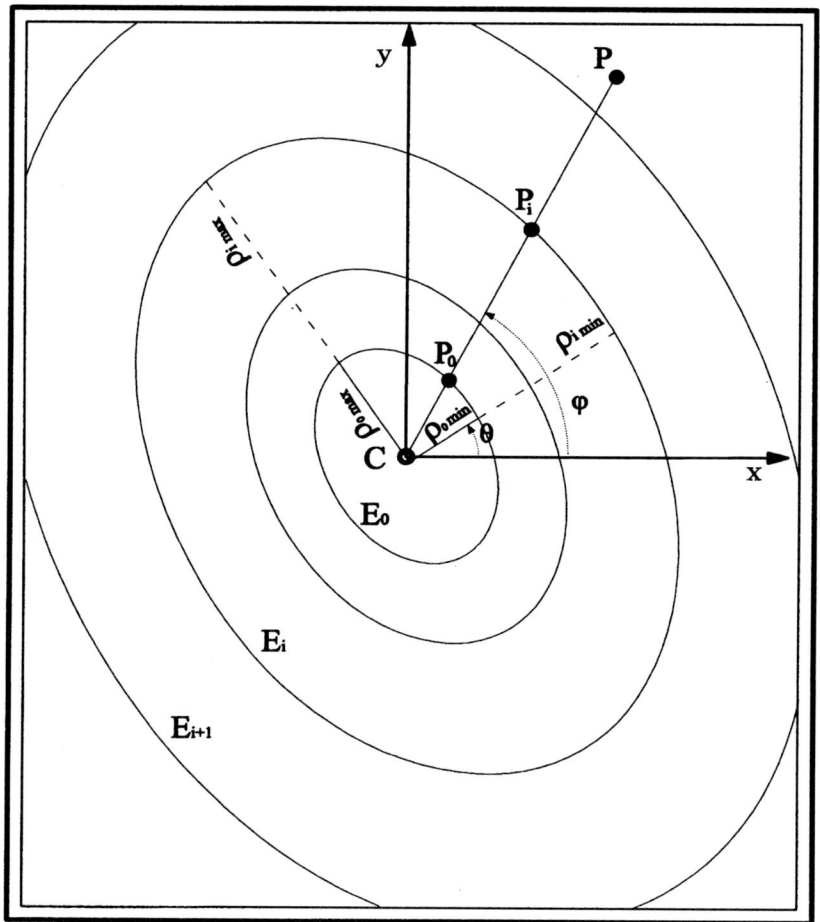

Figure 1
Geometric representation of a virtual intensity distribution.

2. Deduction of the Attenuation Law

With the knowledge of the virtual intensity distribution of a given seismogenic zone (TERAMO et al., 1998), let us consider E_0 the first ellipse and E_i the i-th ellipse, with centers coinciding with the C virtual epicenter, and to which the epicentral intensities I_0 and I_i are respectively associated (Fig. 1).

Moreover, let us denote by ($\rho_{0\ max}$, $\rho_{0\ min}$) and ($\rho_{i\ max}$, $\rho_{i\ min}$) the largest and the smallest semi-axis of E_0 and E_i two ellipses; θ the angle between $\rho_{0\ min}$ (that singles out the maximum attenuation direction) and the x axis of the system of reference $C(x,y)$; φ the angle between the x axis and the line joining the P generic point with the C center of the ellipse; P_0 and P_i the intersections of the semi-straight line CP with the ellipses E_0 and E_i.

Because the ratio between the semi-axes of the ellipses that characterizes the virtual intensity distribution (TERAMO et al., 1998) is equal to:

$$\frac{\rho_{i\ max}}{\rho_{0\ max}} = \frac{\rho_{i\ min}}{\rho_{0\ min}} = K^{i/2} \tag{1}$$

and it is:

$$I_0 - I_i = i, \tag{2}$$

where I_0 is the epicentral intensity; I_i represents the ith intensity level; $i = 0, 1, 2, 3, ...,$ I_{0-1} is the i-th isoseismal line, we obtain, with reference to i-th isoseismal line:

$$CP_i = CP_0 \cdot K^{1/2},$$

with $CP_{0 =} ((\rho_{0\ max}\ \text{sen}\ (\theta - \varphi))^2 + (\rho_{0\ min}\ \cos\ (\theta - \varphi))^2)^{1/2}$.

From which, being: $i = \log(CP_i/CP_0)/\log(k)$, for relation, (2) denoted by d_0 and d_i the CP_0 and CP_i distance, we obtain:

$$I_i = I_0 - \beta \cdot \log(d_i/d_0)/\log(k), \tag{3}$$

β being the attenuation coefficient relative to the CP_i direction.

Relation (3) that characterizes the attenuation law of macroseismic intensity, configured on virtual intensity distribution for seismogenic zones, can be written in the form:

$$I_i = I_0 - \alpha \cdot \log(d_i/d_0), \tag{4}$$

where α is the regional attenuation coefficient with reference to the r direction singled out from the line joining the C virtual epicenter to the P generic point.

It is to be observed that the close correlation between the proposed attenuation law and the virtual intensity distribution shows error levels referable only to a precise evaluation of the observed data which are carried out by filtering procedures (1st and 2nd grade filtered macroseismic planes) (TERMINI et al., 2005). The efficacy of attenuation modelling is therefore strictly connected with the consistency of the virtual intensity distribution that derives from filtered macroseismic planes of earthquakes located inside the seismogenic zone in study, and for which subsequent and stricter configurations are possible.

Finally, it is to be considered that the law provides a modelling of attenuation of an earthquake that configures itself as a characteristic one of a given seismogenic zone, by the corresponding virtual intensity distribution, independently from the epicentral intensity level. Geometrically, this is represented by a set of curves that are shifted along the intensity axis. By the same procedure, and with reference to the *local* virtual intensity distribution, (TERMINI *et al.*, 2005), it is possible to model the attenuation of an earthquake, evaluating the consistency with the observed data.

3. *Analyses and Results*

The proposed attenuation law was tested for different values of epicentral intensity and with reference to thirty-two directions, by *regional* virtual intensity distribution, characteristic of the seismogenic zones individualized by the numbers 71, 72, 74, 75, 78 and 79 in the seismic zonation (SCANDONE *et al.*, 1992) adopted from GNDT (Gruppo Nazionale per la Difesa dai Terremoti, C.N.R., Rome) for the seismic hazard of Italy, in the last version of April 1996 (Fig. 2).

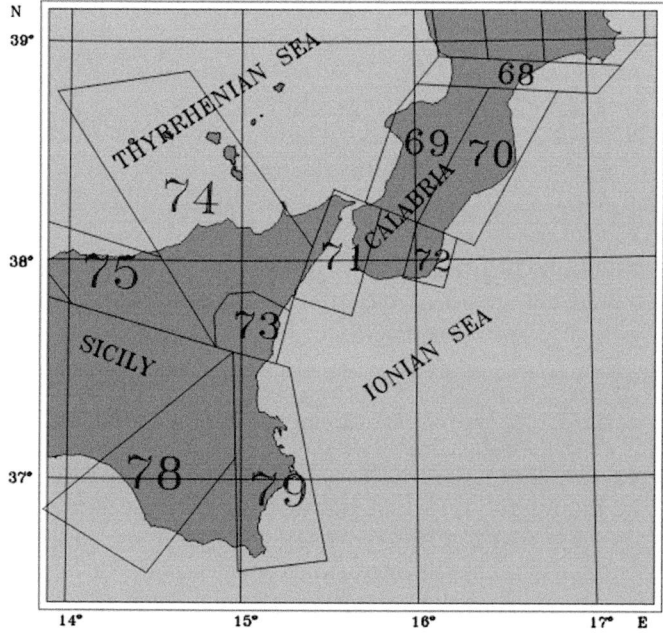

Figure 2
Seismogenic zonation of southern Italy.

Consequently and briefly, only the results relative to an attenuation of the virtual intensity distributions of seismogenic zones 71 and 79 are reported, with reference to the corresponding virtual directions of minimum and maximum attenuation and for different values of epicentral intensity.

All diagrams (Figs. 3 and 4) highlight low error levels and a good consistency with data.

Another significant test was carried out evaluating the intensity attenuation of historical earthquakes by two different approaches: the first one characterizes a *local* type attenuation, deduced namely by parameters obtained by local virtual intensity distribution; the second one, on the contrary, characterizes a *regional* type attenuation, deduced by parameters achieved by regional virtual intensity distribution.

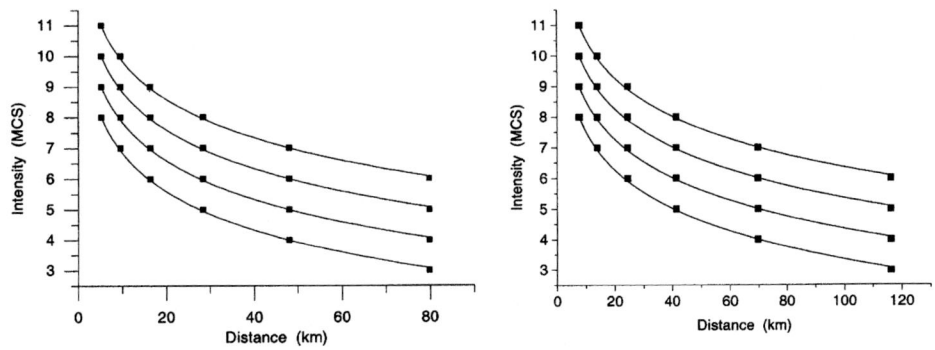

Figure 3
Attenuation of macroseismic intensity for 71 seismogenic zone relative to the minimum (left) and maximum (right) attenuation directions.

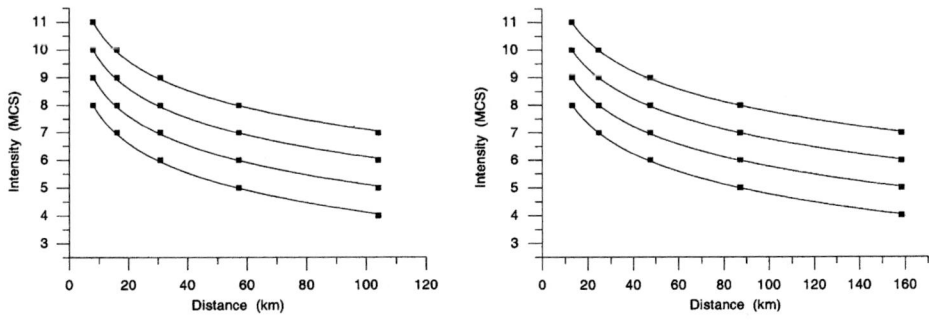

Figure 4
Attenuation of macroseismic intensity for 79 seismogenic zone relative to the minimum (left) and maximum (right) attenuation directions.

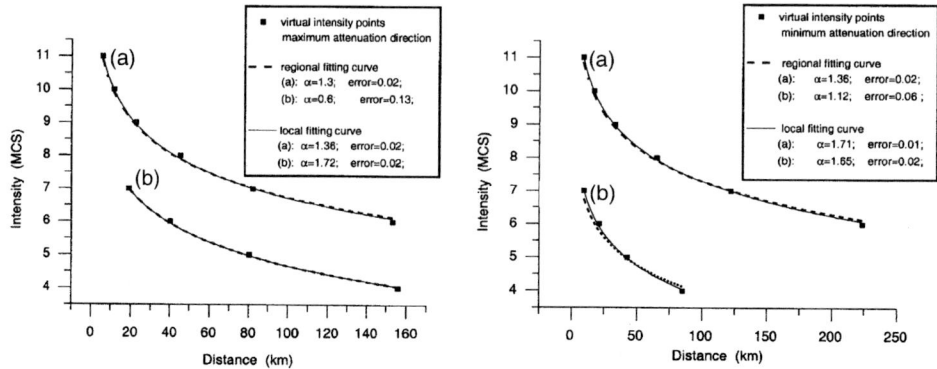

Figure 5
Attenuation of macroseismic intensity of the 28-12-1908 Messina earthquake (a) and the 16-01-1975 Reggio Calabria earthquake (b) with reference to the respective minimum attenuation directions: Regional (full line) and local (hatched line) fitting curves.

For the two seismogenic zones in study, numbers 71 and 79, earthquakes with different values of epicentral intensity were chosen: for zone 71, the earthquakes of 28-12-1908 Messina ($I_0 = XI$ MCS) and of 16-01-1975 Reggio Calabria ($I_0 = VII$ MCS); for seismogenic zone 79, the earthquakes of 11-01-1693 Noto ($I_0 = XI$ MCS) and of 10-12-1542 Siracusa. The corresponding diagrams are reported in Figs. 5 and 6.

From a comparison of two models of attenuation, regional and local ones, no significant differences emerge. This allows therefore, by a good approximation, the *regional* attenuation as a characteristic of the seismogenic zone to be defined with the subsequent possibility of configuring seismic hazard scenarios for earthquakes with different values of epicentral intensity.

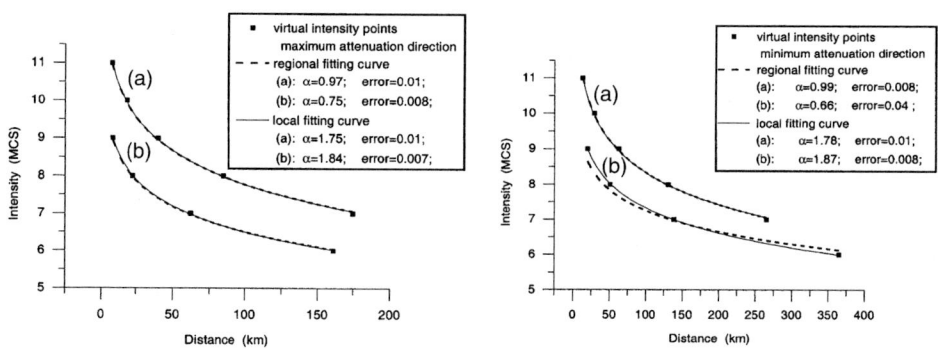

Figure 6
Attenuation of macroseismic intensity of the 11-01-1693 Noto earthquake (a) and the 10-12-1542 Siracusa earthquake (b) with reference to the respective minimum attenuation directions: Regional (full line) and local (hatched line) fitting curves.

Finally, it is to be highlighted that the proposed attenuation law derives from the same vectorial modelling of macroseismic intensity distribution used for the characterization of another anisotropic local attenuation law (TERAMO *et al.*, 1995), that shows the same structure of the relation (4):

$$I_i = I_0 - \alpha \cdot \log(\lambda_i/\lambda_0), \tag{5}$$

where I_i represents the i-th intensity level; I_0 is the epicentral intensity; α is the regional attenuation coefficient; λ_i, λ_0 represent, respectively, the principal moment of inertia of the i-th isoseismal area and the epicentral one, with respect to the virtual directions of minimum and maximum macroseismic intensity attenuation.

From equation (5), by the modelling of virtual intensity distribution, a set of calculations was enucleated, which depicted by the parameters highlighted in the formula (4), makes its use easier.

4. Conclusions

An attenuation law of macroseismic intensity for seismic hazard evaluations was characterized, strictly connected to the virtual intensity distribution for seismogenic zones. The corresponding diagrams of intensity-epicentral distance, relative to a given seismogenic zone and for different values of epicentral intensity, are characteristic of that zone and efficient in defining risk scenarios.

The reliability of such a law in the intensity attenuation modelling of a seismogenic zone derives both from a low error level detected in the corresponding fitting curve analysis, and from the possibility of application to single earthquakes belonging to the same seismogenic zone, by the consistency verification with the observed intensity distribution. The analyses performed on earthquakes belonging to seismogenic zones 71 and 79 located near the Straits of Messina and south-eastern Sicily, confirm the validity of the proposed modelling.

REFERENCES

GRANDORI, G., PEROTTI, F., and TAGLIANI, A. (1987), *On the attenuation of macroseismic intensity with epicenter distance*. Ground motion and Engineering Seismology, (A.S. Camkak, ed.), pp. 581–594, 3rd Int. Conf. On *Soil Dynamics and Earthquake Eng.*, Princeton, USA, 1986 (Elsevier, Amsterdam, 1987).

MAUGERI, M., MOTTA, E., and ZENG, X. (1993), *Attenuation Laws of Seismic Intensity in the Regions of Sicily and Calabria*, Soil Dynam. Earthq. Engin. *12*, 25–35.

PERUZZA, L. (1995), *Macroseismic intensity versus distance: Constraints to the attenuation model*. In (Camkak A.S., Brebbia C.A, eds), *Soil Dynamics and Earthquake Engeneering VII*, Comp. Mech. Publ., Southampton, pp. 215–222.

SCANDONE, P., PATACCA, E., MELETTI, C., BELLATALLA, M., PERILLI, N., and SANTINI, U. (1992), *Struttura geologica, evoluzione cinematica e schema sismotettonico della penisola italiana*. Atti del Convegno del Gruppo Nazionale per la Difesa dai Terremoti, Pisa, 25–27 Giugno 1990. Zonazione e Riclassificazione sismica, *1*, 119–135.

TERAMO, A., STILLITANI., E. and BOTTARI, A. (1995a), *Anisotropic Characterization of Macroseismic Fields,* Natural Hazards *11,* 223–245.

TERAMO, A., STILLITANI, E., and BOTTARI, A. (1995b) *On an Anisotropic Attenuation Law of the Macroseismic Intensity,* Natural Hazards 11, 203–221.

TERAMO, A., TERMINI, D., and STILLITANI, E. (1995c), *La modellazione della distribuzione dell'intensità mediante una ridefinizione del piano macrosismico.* Atti 14° Conv. Ann. CNR- G.N.G.T.S., Roma, 23–25 Ottobre 1995, *1,* 211–216.

TERAMO, A., TERMINI, D., and STILLITANI, E. (1995d), *Sui livelli di affidabilità dei piani macrosismici.* Atti 14° Conv. Ann. CNR- G.N.G.T.S., Roma, 23–25 Ottobre 1995; *1,* 217–221.

TERAMO, A., TERMINI, D., STILLITANI, E., and BOTTARI, A. (1996), *The Determination of the Epicenter by a Vectorial Modelling of Macroseismic Intensity Distribution,* Natural Hazards *13,* 101–117.

TERAMO, A., TERMINI, D., STILLITANI, E., and BOTTARI, A. (1998), *An Anisotropic Modelling for the Determination of the Regional Attenuation Coefficients,* Natural Hazard *17,* 17–30.

TERMINI, D., TERAMO, A., TUVE, T., and BOTTARI, A. (2005), *On the Observed Intensity filtering in the Anisotropic Distribution Modelling of Macroseismic Intensity,* Pure Appl. Geophys.162, 683–697.

(Received February 16, 2001, revised October 10, 2001, accepted December 26, 2001)

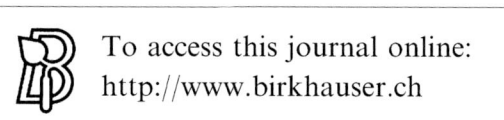

To access this journal online:
http://www.birkhauser.ch

Pure appl. geophys. 162 (2005) 715–727
0033–4553/05/040715–13
DOI 10.1007/s00024-004-2635-x

© Birkhäuser Verlag, Basel, 2005

| **Pure and Applied Geophysics**

Macroseismic Magnitude Evaluation of Earthquakes of Calabria and Northeastern Sicily (Southern Italy)

DOMENICA TERMINI[1], ANTONIO TERAMO[1], and CARLA BOTTARI[1]

Abstract—A comparison of the most used magnitude-intensity relations is carried out, with reference to a spatial window coinciding with Calabria and northeastern Sicily, evaluating their consistency with different data sets taken from several catalogues. *M* values indicated in the catalogues and the corresponding average value have been used choosing the relation to be adopted on the basis of the level of their adaptability to points, rather than prearranged models.

Key words: Macroseismic magnitude, intensity-magnitude relation.

1. Introduction

Over the years, several statistical analyses have been carried out to define a mathematical relation between the epicentral intensity and magnitude (*M* versus *I*) suitable for the determination of the macroseismic magnitude of historical Italian earthquakes. The goal is of an effective interest because the earthquakes for which both the epicentral intensity and instrumental magnitude are known, are a small percentage of all earthquakes of a long Italian seismic history, for which the epicentral intensities are known. The interest in this topic increased in the eighties due to the remarkable contribution of many authors who collaborated in drafting the Catalogue of Italian earthquakes from the year 1000 to 1980 (National Research Council, PF Geodinamica, ed. Postpischl). This high interest was also maintained in the nineties through the acquisition of a better knowledge of the greatest historical earthquakes (CAMASSI and STUCCHI, 1997) and owing to the need to depict an approach for the evaluation of the seismic hazard on the basis of more reliable macroseismic parameters. In short, after first attempts to model empirical relations intensity-magnitude for a set of earthquakes in different Italian regions (MARCELLI and MONTECCHI, 1962; KARNIK, 1969), new modellings are followed on the basis of the analysis of more numerous data sets and of a better quality of the same ones (PERONACI, 1982; TINTI *et al.*, 1986; DI MARO and TERTULLIANI, 1990).

[1] Osservatorio Sismologico, Università di Messina, Via Osservatorio, 4 – 98121 Messina, Italy.
E-mail: dtermini@unime.it

Strong emphasis was given by CAPUTO (1983) to the difficulty in defining correctly a one-to-one relation between the M magnitude and I epicentral intensity; M being a continuous physical quantity, and I, an empirical discrete quantity, describing the effects produced by seismic shaking. Nevertheless, over the years, several relations have been proposed by many authors. Among these we show some of the most utilized ones here:

$$M = 0.481 \cdot I + 1.407 + a \qquad (1)$$

by MARCELLI and MONTECCHI (1962), derived from a set of 83 earthquakes, suitable for $I \leq$ VIII MCS, with a regional term. For Basilicata, northern Calabria and southern Calabria, eastern Sicily, the regional term is equal to -0.02 and 0.04, respectively. For earthquakes with $I \geq$ VIII MCS the same authors opt for a quadratic relation:

$$M = 0.024 \cdot I^2 + 0.206 \cdot I + 2.157. \qquad (2)$$

KARNIK (1969) characterizes a simple linear relation for the Italian territory, through a set of 295 pairs of data (I, M). In particular, for the earthquakes of southern Italy he proposes the relation

$$M = 0.35 \cdot I + 2.1. \qquad (3)$$

CONSOLE *et al.* (1973), through the analysis of 238 seismic events, characterize a relation in which the dependence from the h hypocentral depth is represented in an explicit form. For Calabria-NE Sicily the proposed relation is

$$M = 0.414 \cdot I + 1.105 \cdot log(h) + 0.367. \qquad (4)$$

COSENTINO and LOMBARDO (1980) for 89 earthquakes of Calabria and Sicily propose the relation

$$M = 0.52 \cdot I + 1.35. \qquad (5)$$

PERONACI (1982), using a set of 851 earthquakes, proposes the simple linear dependence of M from I already adopted both from MARCELLI and MONTECCHI (1962) and KARNIK (1969). In particular, for the earthquakes of Calabria he obtains

$$M = 0.654 \cdot I + 0.471. \qquad (6)$$

CAVALLINI and REBEZ (1996) notice a better adaptability of the 195 pairs of values extracted from the GNDT Catalogue (STUCCHI *et al.*, 1993) to a nonlinear relation. In particular they verify that for the interval of values IV$\leq I \leq$ XI MCS the best fit is obtained adopting a sigmoid function

$$M = \{M^{(\mu,\sigma)}[I], M^{(\mu,\sigma)}[I]\} = \mu + 2^{0.5}\mathrm{Erf}^{-1}[\left(\frac{1}{6}\right)I - 1]. \qquad (7)$$

Recently, a review of the strong and moderate earthquakes ($M \geq 4.5$) that affected the southern Apennine (NIKONOV *et al.*, in preparation) proposed again the need to deepen the topic and to define an empirical relation $M = M(I)$, suitable for the determination of the macroseismic magnitude of pre-instrumental epoch earthquakes of this region which is among those in the highest seismic hazard of the entire Italian territory. For a set of 91 earthquakes, originating in a territory wider than the reference one of the NIKONOV catalogue, BOTTARI et al. (1998) fitted (I,M) points through a function of type

$$M = pI^{\alpha} log(I) + q, \tag{8}$$

where p = 0.11, q = 2.90 and α = 1.5 .

Starting with such preliminary remarks, several tests were carried out on pair sets I, M, extracted from the PF Geodinamica, NT4.1 and Nikonov catalogues, respectively, in order to back the choice of the type of relation to adopt for the evaluation of macroseismic magnitude.

2. Tests on Data of Catalogues

All the above indicated M versus I relations, if we exclude the one that represents in an explicit form the dependence from the hypocentral depth (4), are of four types: linear, parabolic, sigmoid and logarithmic. The authors, in proposing them for the evaluation of macroseismic magnitude, have characterized them using data sets generally different.

In a preliminary way to verify the adaptability level of such relations to data, we used them to fit I,M pairs extracted from different catalogues: PF Geodinamica (POSTPISCHL, 1985), NT4.1.1 (STUCCHI et al., 1997), NIKONOV et al.(in preparation).

The epicentral intensities indicated in the three catalogues are all expressed by MCS grades. Differently, magnitudes are not homogeneous: M_L in the Postpischl catalogue, M_s and M_m in the NT4.1 one, M_s in the Nikonov one. Nevertheless, considering that the aim of the analysis is to verify which type of relation better adapts itself to (I, M) coordinate point, the magnitude of each data extracted set is homogeneous: M_L in the Postpischl catalogue, M_s in the NT4.1 and the Nikonov one. In succession, we will omit specifying the type of magnitude and then such a quantity will be indicated without distinction by M.

Three I, M sets have been extracted from the mentioned catalogues, relative to the same time interval (from 1905 to 1980), to the same territorial window (37.5÷40° N; 14÷17°E) (Fig .1), to the same epicentral intensities interval: $V \leq I \leq XI$ MCS and for hypocentral depth h < 40 km. The adopted time interval is the longest possible compatible with the extent of the three considered catalogues. The chosen territorial window is referable to the need to develop the analysis only with reference to the earthquakes that took place inside the region considered by the Nikonov catalogue. Such a region, consistent with the seismic zonation of SCANDONE et al. (1992),

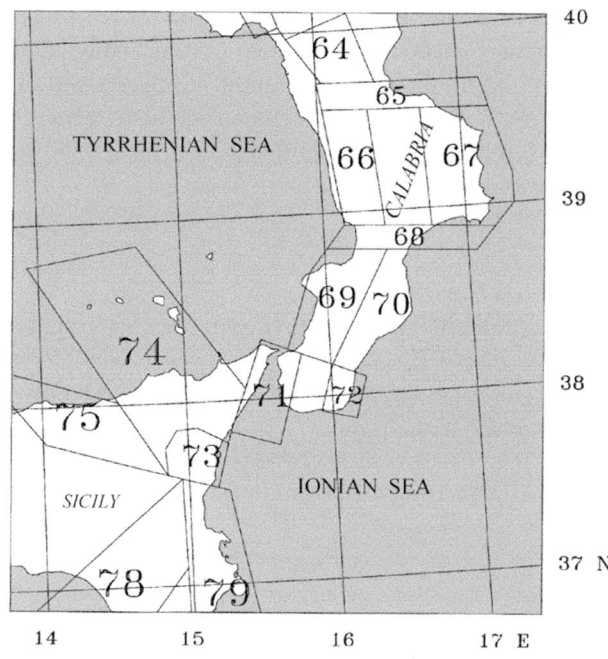

Figure 1
Seismogenic zonation of southern Italy.

includes, proceeding from North to South, the seismic zones numbers 64, 65, 66, 67, 68, 69, 70, 71 and 72.

The interval $V \leq I \leq XI$ MCS, adopted for the epicentral intensities, is referable to intensity values concerning permanent seismic effects (from small cracks to collapse). This improves the reliability level of data on which to carry out regressions and directs the analysis towards the refining of parameters of specific interest for the evaluation of seismic hazard.

The three data sets I, M extracted from the three catalogues considering the above-mentioned limits and excluding earthquakes that took place in the Aetnean area (BOTTARI *et al.*, 1998; AZZARO and BARBANO, 1997), have different numerical weights: 35 for the Postpischl catalogue, 87 for the NT4.1 one and 31 for the Nikonov one. With reference to the NT4.1 catalogue, which gives the most numerous set of earthquakes, the Nikonov catalogue provides less than half considering that owing to the $M \geq 4.5$ threshold it does not include those earthquakes with a magnitude smaller than 4.5. All three catalogues show point densities considerably greater in the interval of average and average-high epicentral intensities ($V \leq I \leq VIII$ MCS) as it regards the high ones ($I > VIII$ MCS).

Relative to data regression I, M adopting a sigmoid function, in order to fit the magnitude M as a dependent variable and the epicentral intensity as an independent variable, we employ the relationship

$$I = I_{\max} \mathrm{Erf}\{[(M - \mu)/\sigma]\} + 1. \tag{9}$$

The +1 value is the minimum intensity to be considered (instead of 0 as assumed by CAVALLINI and REBEZ, 1996) according to the MCS scale that starts from 1 (and not zero). We propose measuring this stochastic component using the normal function N (μ,σ), μ and σ being the mean and the standard deviation of the magnitude values of earthquakes in the data set,

$$M = \mu + \sigma \mathrm{Erf}^{-1}\{[(I - 1)/I_{\max}\}]. \tag{10}$$

$I_{.\max}$ is a parameter introduced to fit the intensity values to the data interval, taking into account that we need to add a unit to start the intensity scale with a non zero value.

To derive M from I values the inverse error function (Erf $^{-1}$) must follow the condition

$$0 \le (I - 1)/I_{\max} \le 1 \tag{11}$$

as the variable in Erf $^{-1}$ must be a probability value. This condition is valid only when $I \le I_{\max}$ and it imposes a limit to the estimation of M.

The detailed results of the regressions are shown in Table 1.

In succession, in order for carry out regressions on an equal number of (I, M) points, to each value of epicentral intensity the corresponding M average value is associated, and calculated for every set of M values given by each catalogue. The

Table 1

Different magnitude-intensity relations for M values given by catalogues.

CATALOGUE	Linear relation	Parabolic relation	Sigmoid relation	Logarithmic rel.
POSTPISCHL (1985)	$a = 1{,}560 \pm 0{,}474$ $b = 0{,}468 \pm 0{,}072$	$a = 5{,}478 \pm 1{,}986$ $b = -0{,}653 \pm 0{,}558$ $c = 0{,}076 \pm 0{,}037$	$\mu = 4{,}579$ $\sigma = 1{,}858$	$p = 3{,}173 \pm 0{,}226$ $q = 0{,}101 \pm 0{,}015$
	Points: 35 $R^2 = 0{,}626$ SD = 0,626	Points: 35 $R^2 = 0{,}613$ SD = 0,599	Points: 35 $R^2 = 0{,}582$ SD = 0,613	Points: 35 $R^2 = 0{,}593$ SD = 0,604
NT4.1 (1996)	$a = 1{,}059 \pm 0{,}071$ $b = 0{,}562 \pm 0{,}011$	$a = 1{,}598 \pm 0{,}314$ $b = 0{,}413 \pm 0{,}086$ $c = 0{,}010 \pm 0{,}006$	$\mu = 4{,}689$ $\sigma = 2{,}140$	$p = 3{,}092 \pm 0{,}034$ $q = 0{,}116 \pm 0{,}002$
	Points: 87 $R^2 = 0{,}984$ SD = 0,110	Points: 87 $R^2 = 0{,}969$ SD = 0,108	Points: 87 $R^2 = 0{,}960$ SD = 0,122	Points: 87 $R^2 = 0{,}963$ SD = 0,117
NIKONOV (1997)	$a = 1{,}086 \pm 0{,}279$ $b = 0{,}546 \pm 0{,}037$	$a = 3{,}365 \pm 1{,}107$ $b = -0{,}047 \pm 0{,}282$ $c = 0{,}037 \pm 0{,}017$	$\mu = 4{,}667$ $\sigma = 2{,}067$	$p = 3{,}151 \pm 0{,}131$ $q = 0{,}110 \pm 0{,}069$
	Points: 31 $R^2 = 0{,}939$ SD = 0,263	Points: 31 $R^2 = 0{,}898$ SD = 0,248	Points: 31 $R^2 = 0{,}895$ SD = 0,247	Points: 31 $R^2 = 0{,}894$ SD = 0,248

Table 2

Different magnitude-intensity relations for average M values

CATALOGUE	Linear relation	Parabolic relation	Sigmoid relation	Logarithmic rel.
POSTPISCHL (1985)	$a = 0,966 \pm 0,487$ $b = 0,553 \pm 0,062$	$a = 4,201 \pm 1,902$ $b = -0,309 \pm 0,496$ $c = 0,054 \pm 0,031$	$\mu = 4,567$ $\sigma = 2,068$	$p = 3,012 \pm 0,234$ $q = 0,112 \pm 0,011$
	Points: 11 $R^2 = 0,948$ SD = 0,370	Points: 11 $R^2 = 0,926$ SD = 0,333	Points: 11 $R^2 = 0,907$ SD = 0,353	Points: 11 $R^2 = 0,921$ SD = 0,325
NT4.1 (1996)	$a = 1,028 \pm 0,077$ $b = 0,567 \pm 0,010$	$a = 1,493 \pm 0,276$ $b = 0,443 \pm 0,072$ $c = 0,008 \pm 0,004$	$\mu = 4,711$ $\sigma = 2,047$	$p = 3,159 \pm 0,076$ $q = 0,112 \pm 0,004$
	Points: 9 $R^2 = 0,999$ SD = 0,054	Points: 9 $R^2 = 0,998$ SD = 0,047	Points: 9 $R^2 = 0,989$ SD = 0,116	Points: 9 $R^2 = 0,991$ SD = 0,105
NIKONOV (1997)	$a = 1,131 \pm 0,351$ $b = 0,547 \pm 0,044$	$a = 4,398 \pm 1,120$ $b = -0,316 \pm 0,290$ $c = 0,054 \pm 0,018$	$\mu = 4,640$ $\sigma = 2,0971$	$p = 3,108 \pm 0,122$ $q = 0,109 \pm 0,006$
	Points: 11 $R^2 = 0,972$ SD = 0,274	Points: 11 $R^2 = 0,974$ SD = 0,199	Points: 11 $R^2 = 0,967$ SD = 0,213	Points: 11 $R^2 = 0,978$ SD = 0,173

results of this second series of fit are shown in Table 2. The analysis of results of this Table shows standard deviation values of residuals, relative to regression carried out on data in the Postpischl and Nikonov catalogues, systematically decreasing adopting sequentially the linear, parabolic and logarithmic relations. For the NT4.1 catalogue the best fit of points of coordinates *I, M* is obtained by adopting the parabolic relation.

The comparison of the results of regressions indicated in Table 2 allows some evaluations. Relative to fits on data of the Postpischl catalogue, the standard deviation values are systematically higher for all the adopted relations. Moreover, the standard deviation values relative to parabolic, sigmoid and logarithmic relations do not show appreciable differences (0.599, 0.613 and 0.604 respectively).

The data set extracted from NT4.1 catalogue is that for which systematically we have the smallest values of standard deviation and for which the influence of the type of regression is minimum ($0.108 \leq SD \leq 0.122$). We have similar results from regressions of *I, M* point sets extracted from the Nikonov catalogue, although in this case the standard deviations are intermediate to those obtained for the Postpischl and NT4.1 catalogues ($0.245 \leq SD \leq 0.270$). A better adaptability to data, even if minimal, is obtained by adopting the sigmoid relation. With respect to this, the calculated values of μ and σ coefficients are very close to those obtained by CAVALLINI and REBEZ (1996) for the earthquakes of the entire national territory ($\mu = 4.57$ and $\sigma = 2.57$), however for the earthquakes of southern Apennine and northeastern Sicily considered here, the standard deviation is appreciably lower: 0.122 for the NT4.1 catalogue data, and 0.309 for the Nikonov one instead of 0.492.

Figures 2 to 5 show the logarithmic and sigmoid relations, applied to I,M values of NT4.1 and Nikonov catalogues. Dashed lines indicate the confidence interval.

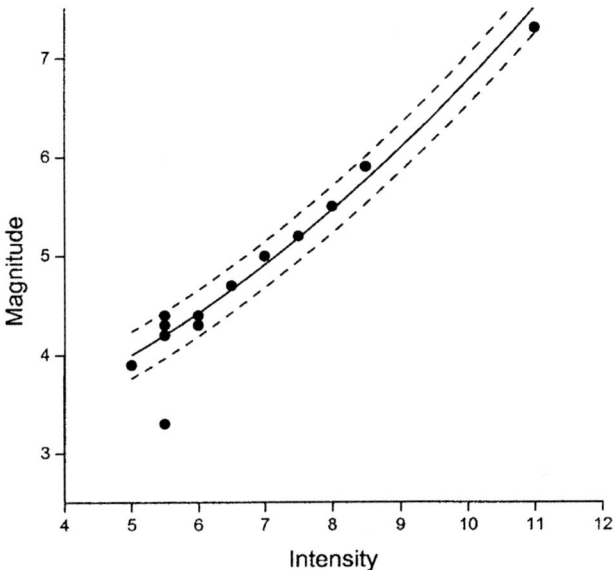

Figure 2
Logarithmic relation applied to the NT4.1 catalogue data.

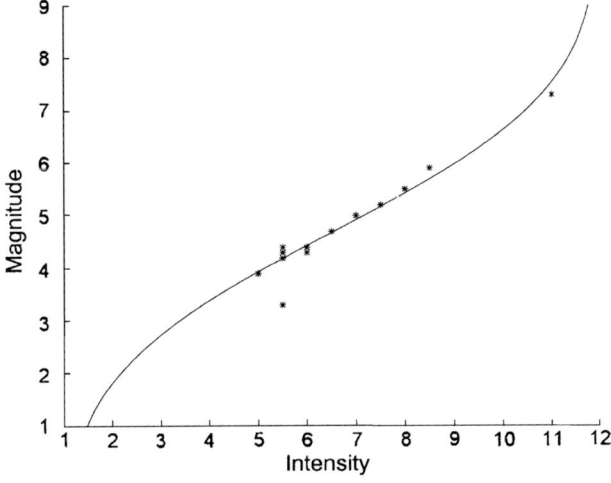

Figure 3
Sigmoid relation applied to the NT4.1 catalogue data.

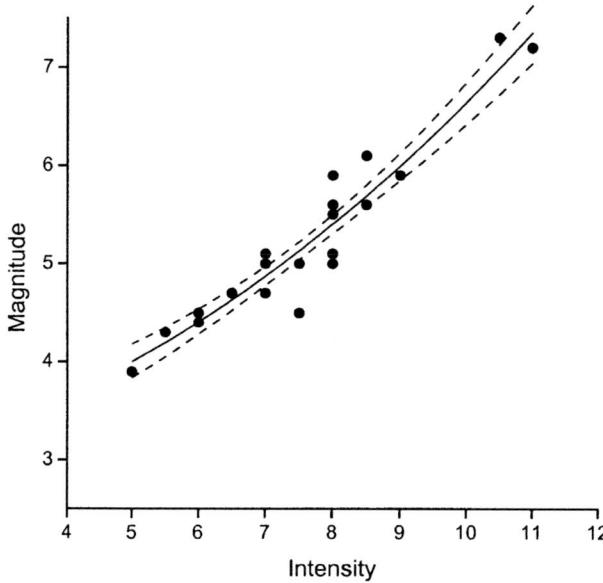

Figure 4
Logarithmic relation applied to the Nikonov catalogue data.

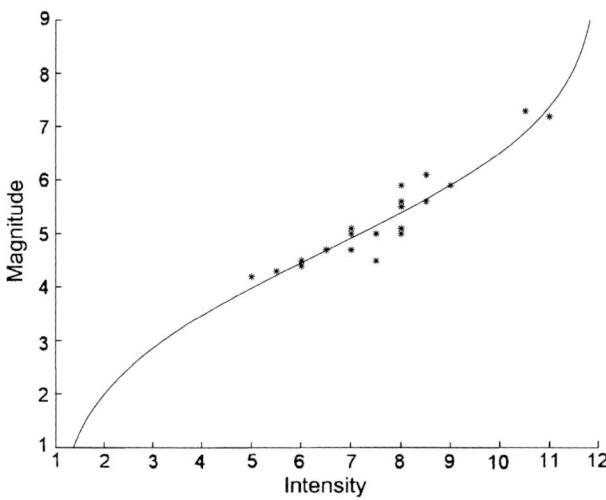

Figure 5
Sigmoid relation applied to the Nikonov catalogue data.

We have already observed that the three tests which have been carried out working on *I, M* point sets, are rather different also for numbers. In order to carry out the regressions of data sets equally (or almost) numerous, all fits have been carried out again, giving each of I values the corresponding average value of M, calculated on the entire magnitude set given by the catalogue for each value of I. Such a procedure offers advantages as a strong reduction of data noise (Sibol *et al.*,

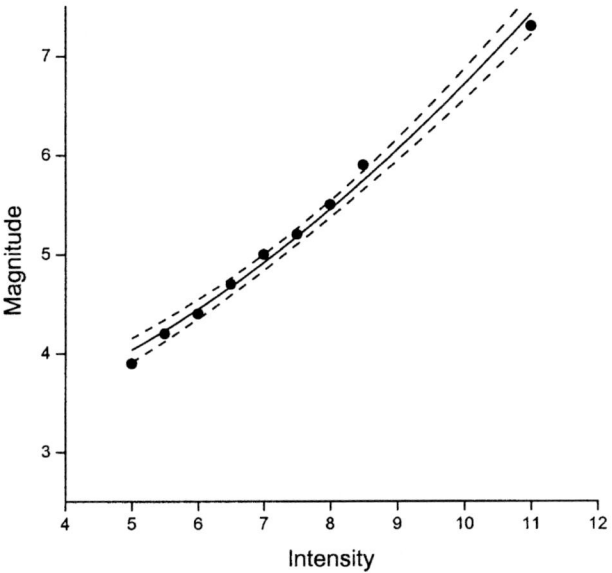

Figure 6
Logarithmic relation applied to the NT4.1 catalogue data with average values of magnitude.

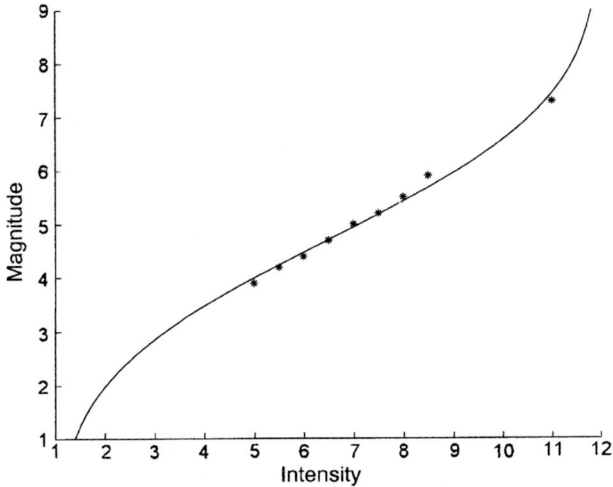

Figure 7
Sigmoid relation applied to the NT4.1 catalogue data with average values of magnitude.

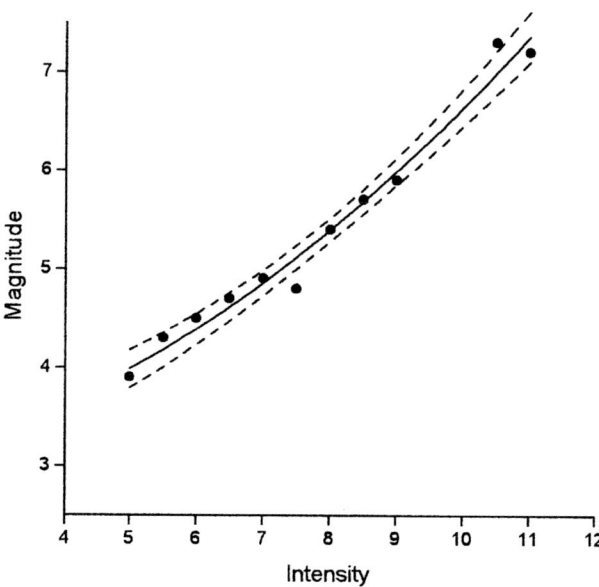

Figure 8
Logarithmic relation applied to the Nikonov catalogue data with average values of magnitude).

1987; CAVALLINI and REBEZ, 1996). The analysis of the results of regressions indicated in Table 3 shows a better adaptability of data to the logarithmic relation both for the Postpischl and the Nikonov catalogues. For the NT4.1 catalogue, the smallest standard deviation value is obtained by adopting the parabolic relation.

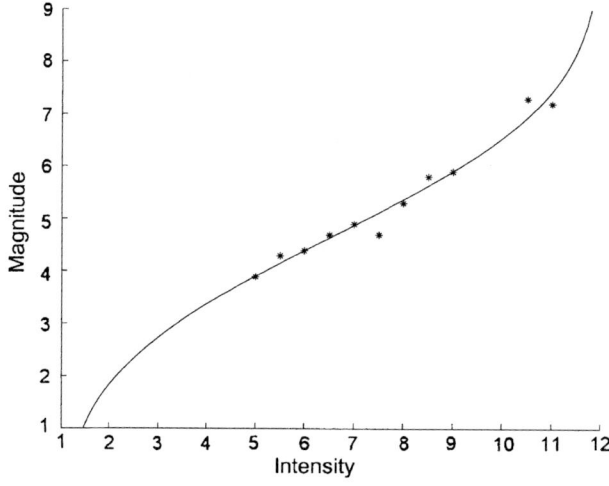

Figure 9
Sigmoid relation applied to the Nikonov catalogue data with average values of magnitude.

It is to be pointed out that the regressions carried out both on full sets and the reduced ones (introducing averaged values of M) show a smaller variability of coefficients adopting the sigmoid relation rather than the linear and the parabolic one.

Figures 6–9 show the logarithmic and the sigmoid relation applied to average values of magnitude corresponding to each epicentral intensity value. Dashed lines indicate the confidence interval.

3. Results and Conclusions

The choice of the type of relation with which to carry out the regression of points (I, M) shows difficulties that caused criticism by different authors. The main objection is relative to the use of correlating the I variable (epicentral intensity), empirical quantity, discrete and essentially qualitative, to M (magnitude), quantitative and continuous quantity, through a one-to-one relation (CAPUTO, 1983). Nevertheless, the practical reasons on the basis of different relations $M = M(I)$, proposed over the years by different authors, have been prevalent over the difficulties of methodological order. In particular, even if it is generally known that the MCS intensity scale does not define how much a given grade is distant from the one that precedes it or from the one that follows it, being a typically ordinal scale, the use of representing I indicating on a Cartesian axis the grades of the scale spaced from all equal intervals is established. Even with such straining and ambiguities the aim is to convert the epicentral intensities of earthquakes of pre-instrumental epoch into M prevailed, and particularly, those of average-high and high intensity. Pragmatically, sharing the priority that such an objective has for the analysis of historical seismicity and specifically for the evaluations of seismic hazard, the choice of the $M = M(I)$ relation is to be evaluated on the basis of the level of adaptability to the points (I, M) rather than with reference to pre-arranged models.

The tests carried out on data of the most common catalogues of Italian earthquakes (POSTPISCHL, 1985 and NT4.1, 1997), with reference to the area including Calabria and northeastern Sicily and for a hypocentral depth $\leq$ 40 km, show standard deviation values, relative to the rather equalized four adopted regression relations. The data set extracted from NT4.1 catalogue shows the smallest standard deviation values. It is to be noticed moreover that for different catalogues and full or reduced data sets, the regression coefficients obtained for sigmoid or logarithmic relation are more steady than the ones relative to the linear and parabolic relations.

Relative to I, M data extracted from the Nikonov catalogue, a better adaptability is observed for the sigmoid and logarithmic relations. The first one is preferred for the regression of a full data set; the second one fits data with average magnitude values better. Considering that the last approach makes the frequency of data for the whole interval of epicentral intensities uniform, the regression according to the relation $M = I^{1,5}$ p log I + q is preferable to the others. On the other hand, considering that the standard deviation values of the sigmoid and logarithmic relations are in any case lower than the average uncertainties ($\pm$ 0.3) of M values, the choice of the logarithmic or sigmoid relations does not lead to significant differences.

In conclusion, the tests carried out indicate that the best conditions of adaptability of the I, M points are obtained through the logarithmic relation (8) and the sigmoid relation (10). This can support their use for the determination of the macroseismic magnitude of crustal earthquakes in the Nikonov catalogue, with an epicentral intensity V $\leq I \leq$ XI MCS, which occurred before the XX century in Calabria and northeastern Sicily.

REFERENCES

AZZARO, R. and BARBANO, M.S. (1997), *Intensity-magnitude Relationship for the Mount Etna Area (Sicily)*, Acta Vulcanologica 9, 15–21.

BOTTARI, A., NIKONOV, A., TERAMO, A., and TERMINI, D. (1998), *Relazione Magnitudo-intensità per i terremoti della Regione dell'Appennino Meridionale* .Società Geologica Italiana. Atti del 79° Congresso Nazionale. Palermo 21–23 Sett. 1998.

CAMASSI, R. and STUCCHI, M. (1997), *NT4., 1 un catalogo parametrico di terremoti di area italiana al di sopra della soglia del danno*, GNDT, Internal report, 84 pp.

CAPUTO, M. (1983), *Are there One-to-one Relationships between Magnitude, Moment, Intensity and Ground Acceleration ?* Geophys. J.R. Astron. Soc. *72*, 83–92.

CAVALLINI, F. and REBEZ, A. (1996), *Representing Earthquake Intensity-Magnitude Relationship with a Nonlinear Function*, Bull. Seismol. Soc. Am. *86, 1A*, 73–78.

CONSOLE, R., PERONACI, M., and SONAGLIA, A. (1973), *Relazione sui fenomeni sismici dell'Anconetano (1972)*. Ann. Geofis. Suppl. *265*, 148 pp.

COSENTINO, M. and LOMBARDO, G. (1980), *Correlation between the Modified Mercalli Scale and the Medvedev-Sponheuer-Karnik Scale for Earthquakes of Sicily and Calabria*, Boll. Geofis. Teor. Appl. *85*, 29–37.

DI MARO, R., and TERTULLIANI, A. (1990), *The Relation between Intensity and Magnitude for Italian Earthquakes*. Pure Appl. Geophys. *132*(4), 711–718.

KARNIK, V. *The seismicity of the European Area*, (Reidel, Dordrecht ,1969).

MARCELLI, L., and MONTECCHI, A. (1962), *Contributi per uno studio sulla sismicità dell'Italia*, Ann. Geof. *15*, 160–175.

NIKONOV, A., BOTTARI, A., and TERAMO, A. (in preparation). *Catalogne of strong and moderate (M $\geq$ 4.5 mainly) crustal earthquakes of South Calabria and Northeastern Sicily (37°-40°N; 14°-17°E) for the period 461 B.C. and 1996.*

PERONACI, M. (1982), *Intensity- magnitude Relationships for Italian Regions*, Boll. Geofis. Teor. Appl. *94*, 121–128.

POSTPISCHL, D. (ed.) (1985), *Catalogo dei terremoti italiani dall'anno 1000 al 1980*. Consiglio Nazionale delle Ricerche. Progetto Finalizzato Geodinamica, Roma, 239 pp.

SCANDONE, P., PATACCA, E., MELETTI, C., BELLATALLA, M., PERILLI, N., and SANTINI, U. (1992), *Struttura geologica, evoluzione cinematica e schema sismotettonico della Penisola Italiana*. Atti del convegno GNDT, Pisa 25–27 Giugno 1990, *1*, 119–135. Edizione Ambiente Bologna.

SIBOL, M.S., BOLLINGER, A., and BIRCH, J.B. (1987), *Estimation of Magnitudes in Central and Eastern America Using Intensity and Felt Area*, Bull. Seismol. Soc. Am. *77*, 165–1654.

STUCCHI, M., CAMASSI, R., and MONACHESI, G. (1993), *The working catalogue of G.N.D.T. National Research Council (C.N.R.), Natioonal Group for Defence Against Earthquakes (G.N.D.T.) Working Group 1.2.2 "Macrosismica",* Internal Report, 89 pp. (in Italian).

TINTI, S., VITTORI, T., and MULARGIA, F. (1986), *Regional Intensity-magnitude Relationships for the Italia Territory,* Tectonophysics *127,* 129–154.

(Received February 2, 2001, accepted December 12, 2001)

 To access this journal online:
http://www.birkhauser.ch

Pure appl. geophys. 162 (2005) 729–737
0033–4553/05/040729–9
DOI 10.1007/s00024-004-2636-9

❙ Pure and Applied Geophysics

A Magnitude-Felt Area Relation in the Evaluation of the Magnitude of Historical Earthquakes

Domenica Termini[1], Antonio Teramo[1], and Giuliana Arrigo[1]

Abstract — An evaluation of the magnitude of historical earthquakes is proposed, through an empirical relation based on a felt area of historical earthquakes derived from a vectorial modelling of macroseismic intensity distribution.

Key words: Historical earthquakes, virtual intensity distribution, felt area, magnitude.

1. Introduction

The determination of the magnitude of historical earthquakes represents a problem of relevant importance in seismology. Among the various authors who have proposed relations for magnitude evaluation, NUTTLI and ZOLLWEG (1974) and SIBOL et al., (1987) characterized specific relations based on epicentral intensity and felt area, through different ways and characteristics, main estimators for measuring historical earthquakes for which the observed intensity distribution was available. The relation proposed by Sibol was also used recently by Gasperini and Ferrari for the determination of the magnitude of strong Italian earthquakes (BOSCHI et al., 1997, 2000):

$$M = a + bI_0^2 + c \log^2(FA), \tag{1}$$

where, I_0 and FA represent the epicentral intensity and felt area, respectively.

Yet, it is to be highlighted that through such a relation it is impossible to estimate magnitude in the case in which sufficient data for the evaluation of the felt area is not available, or for seismic events that have an intensity greater than a VIII-IX grade. In such a case relations based on the nearest neighboring criteria are used, through which the magnitude is calculated as an average of the known magnitude of earthquakes that have the same weighted epicentral intensities with an inverse function of epicentral distance. Relations with only epicentral intensity are used too,

---
[1] Osservatorio Sismologico, Università di Messina, Via Osservatorio, 4 - 98121 Messina, Italy.
E-mail: dtermini@unime.it

such as the Rebèz and Stucchi relation (1996), used for the NT4.1.1 catalogue (CAMASSI and STUCCHI, 1998).

It is further considered that through such a relation it is not possible to establish any correlation between observed intensity data and the felt area in terms of structural framework geometry of the seismogenic zone and local geomorphological conditions. Consequently our aim is to achieve the characterization of an empirical relation that allows the magnitude determination through a felt area deduced from the observed intensity distribution of the earthquake in study, for which it is possible to carry out a specific verification of consistency with the structural framework of the seismogenic zone and local geomorphological conditions.

The proposed magnitude-felt area relation is based on a procedure that, starting with methodological approaches taken from a vectorial intensity distribution, allows for each intensity level and even in cases in which a reduced number of observed intensity points are available an evaluation of a felt area that is consistent with a structural framework of the seismogenic zone and with an eventual filtering of the observed intensity levels.

Such a relation was tested on all strong historical and recent earthquakes which occurred in Calabria and Sicily (southern Italy) highlighting, in some cases, inconsistencies in magnitude evaluation with respect to catalogue data.

2. The Magnitude-felt Area Relation

The determination of the felt area geometry of the single earthquake to be used in magnitude evaluations, even in the case of a remarkable lack of data, represents the more representative element of the proposed procedure, considering that it has to be consistent with the structural framework of the seismogenic zone in which the earthquake falls and with observed intensity distribution, supplemented by a data filtering to reduce the weight of evaluation errors, or local amplifications, referable to local geological conditions.

The study starts with research conducted by TERAMO *et al.* (1995, 1996, 1998) and TERMINI *et al.* (2004a, b, c) on macroseismic intensity modelling of historical earthquakes. The base of such an approach is represented by the construction of the macroseismic plane (set of areolas built around the observed intensity points), that allows specific evaluations on the reliability of each observed intensity point, highlighting eventual amplifications of intensity referable to different local geological conditions or to evaluation errors. Such macroseismic planes are subsequently reshaped through the application of suitable filters (TERMINI *et al.*, 2005a), characterizing in such a way, two different types of filtered macroseismic planes: the *1-st level* ones, through which the approximation of observed intensity half degrees to the greater and lower integer ones is carried out; the *2-nd level* ones, that allow an analysis of surface distribution of integer observed intensity levels of a given earthquake, individualizing eventual anomalies that, can further be highlighted by a study in depth of surveys *in situ*.

With reference to these modellings a vectorial distribution of observed intensity was characterized (TERAMO *et al.*, 1995), which allows an analytical determination of the epicenter and principal attenuation directions of the earthquake in study (TERAMO *et al.*, 1996), as well as the individualization of a *virtual intensity distribution* that is characteristic both for a single earthquake (*local*) (TERMINI *et al*, 2005a) and for a seismogenic zone (*regional*), considering all earthquakes that fall within it (TERAMO *et al.*, 1998).

In particular, the local virtual intensity distribution is individualized by a set of similar ellipses with the center coinciding with the calculated epicenter of the earthquake considered, and axes coinciding with the principal attenuation directions, analytically calculated. The reliability of such a modelling is determined through three different check levels with the local macroseismic parameters: the calculated epicenter (coinciding with the ellipses center) will have to be very close to the barycenter of the maximum shaking area of the earthquake; the major axis direction of the ellipses (which individualizes the minimum attenuation direction) will have to be consistent with the structural framework of the seismogenic zone; each ellipsis, to which a decreasing intensity level is associated, starting with the epicentral one that defines the maximum shaking area of the earthquake, must represent, moreover, a suitable envelope of the homonymous observed intensity points.

The local virtual macroseismic intensity distribution, owing to the adopted procedure for its deduction, depicts an intensity distribution consistent both with the tectonic trend of the seismogenic zone and with the observed intensity distribution, as well as with the anisotropic attenuation of intensity (TERMINI *et al.*, 2005b) that allows the intensity attenuation to be characterized, up to the lowest considered levels, even with the lack of corresponding observed data. It is to be considered that, in case of a remarkable lack of observed data (earthquakes with two or three intensity levels), the local virtual intensity distribution of the earthquake in study can be reshaped with reference to regional parameters, relative to the seismogenic zone in which the same seismic event falls, checking the consistency with local macroseismic parameters.

The local virtual intensity distribution allows therefore the shaking area of the earthquake in study to be characterized efficiently. Considering furthermore that the greatest earthquakes exhibit high epicentral intensity levels, it is proposed that the felt area of each earthquake through the ellipses of the corresponding local virtual intensity distribution associated to the relative intensity levels is individualized, starting with IV MCS up to I_0 epicentral intensity.

In Figure 1, as an example, the felt area of the 28-12-1908 Messina earthquake is shown.

Knowing the felt area of the earthquake in study, the determination of magnitude is effected, starting with its consistency of the macroseismic intensity attenuation with

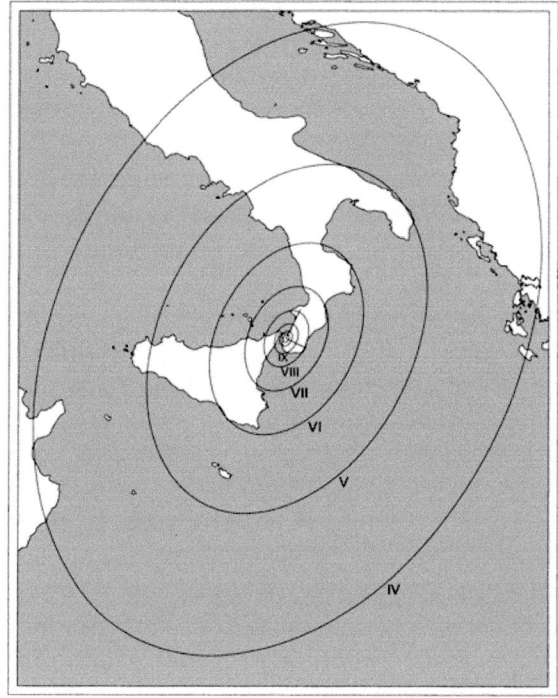

Figure 1
Felt area of the 28-12-1908 Messina earthquake.

the geometry of the ellipses of virtual intensity distribution (TERMINI *et al.*, 2005b), through the characterization of a correspondence between magnitude values with a λ function of A_i ellipses extent of the felt area and the corresponding I_i intensity levels:

$$M = a\lambda(A_i, I_i),\qquad(2)$$

where α is a proportion coefficient.

With reference to crustal earthquakes with foci very near to each other, for the λ function a proportionality of the type can be depicted:

$$\lambda = \ln\sum_i (A_i \cdot I_i^2),\qquad(3)$$

where A_i and I_i represent the extent in km^2 of isoseismal areas of local intensity distribution and the corresponding intensity levels, determined for $i = I_0, \dots , $ IV MCS, respectively (Fig 1). Since, substituting in (2), the following is obtained

$$M = \alpha\ln\sum_I (A_i I_i^2).\qquad(4)$$

Table 1

Comparison between M magnitude values taken from catalogues and M_c calculated ones through the proposed relation

Epicentral zone	Date	SZ	I_0		M		F.A. (10^5 km^2)	M_c
			NT4.1.1	BOSCHI	NT4.1.1	BOSCHI		
Sicilia Orientale	04-02-1169	79	XI	X	7.3	6.5	39	7.3
Sicilia Orientale	11-01-1693	79	XI	XI	7.0	7.4	42	7,3
Calabria Meridionale	05-02-1783	69	XI	XI	7.3	6.9	2.50	6,2
Calabria Centrale	28-03-1783	68	XI	XI	6.7	6.9	49	7,3
Madonie	08-09-1818	75	VII-VIII	VII-VIII	5.2	5.3	0.49	5,5
Bagnara Calabra	16-11-1894	69	IX	IX	5.9	6.2	1.9	6,0
Golfo di S. Eufemia	08-09-1905	69	X-XI	X-XI	7.5	6.8	13	6,8
Ferruzzano	23-10-1907	72	IX	VIII-IX	5.9	6.0	0.20	6,1
Novara di Sicilia	10-12-1908	74	VII	-	5.0	-	0.29	5,3
Messina	28-12-1908	71	XI	XI	7.3	7.1	8.4	6.7
Mar Ionio	11-05-1947	75	IX	VIII	5.6	5.9	3.3	6,2
Sperlinga	31-10-1967	70	VIII	VIII	5.1	5.7	1.1	5,8
Reggio Calabria	16-01-1975	71	VII-VIII	VII-VIII	4.5	5.4	0.27	5,3
Brancaleone	11-03-1978	72	VIII	VIII	5.0	5.5	0.24	5,2
Golfo di Patti	15-04-1978	74	VIII	VIII-IX	6.1	5.8	1.1	5,8

3. Analyses and Results

The reliability of such a relation was checked on a set of historical earthquakes that occurred in different seismogenic zones of Calabria and Sicily (southern Italy), designing previously local virtual intensity distribution of each earthquake and determining subsequently the corresponding value of magnitude through the proposed empirical relation.

In Table 1, for the earthquakes in study, which occurred in different seismogenic zones (S.Z.) of southern Calabria and eastern Sicily, the full extent of the felt area (F.A.) is shown, determined through the local virtual intensity distribution of each seismic event, with reference to foreseen observed intensity levels I_0-IV MCS, together with their I_0 epicentral intensity and M magnitude values taken from the NT4.1.1 catalogue, in the last version of March 1998, and the BOSCHI *et al.* catalogue (2000).

Determining the λ function for all the earthquakes, through two fits with the corresponding values of M magnitude indicated in the two catalogues (Fig. 2), two different values of α coefficient have been calculated for the relation (4). We have chosen $\alpha = 0,40$ that shows the lowest residuals with respect to the more recent earthquakes for which instrumental data are available (after the 28-12-1908 earthquake).

The M_c magnitude values calculated with the proposed relation:

$$M = 0.40 \ln \sum_I (A_I \cdot I_i^2) \tag{5}$$

are compared in Table 1 with those taken from the above-mentioned catalogues.

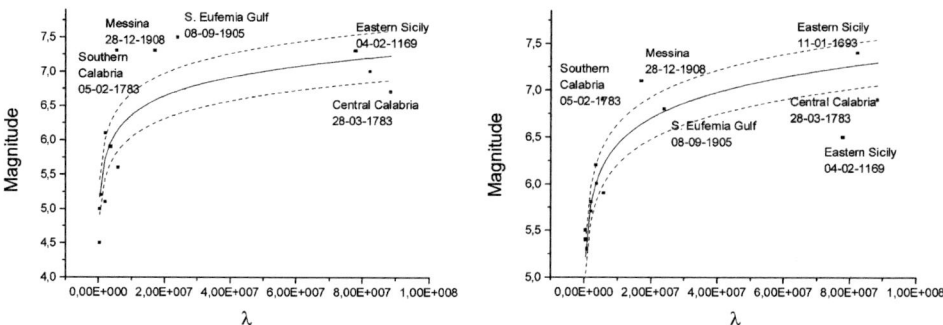

Figure 2

Proposed magnitude-felt area relation diagrams with respect to magnitude values taken from NT4.1.1 catalogue (left) and BOSCHI *et al.* one (right).

The comparisons carried out highlight, in general, a good consistency between different values of magnitude, with the correspondent extent of ellipses of the local virtual intensity distributions. The fit errors, calculated with respect to magnitude values taken from the NT4.1.1 and BOSCHI *et al.* Catalogues, vary from 0,006 up to 0,009 and the α coefficient has in both cases the same value equal to 0,40.

With greater detail, for the five events with the epicentral intensity $I_0 = XI$ MCS which occurred in Calabria and Sicily (southern Italy), evaluations of magnitude taken from the NT4.1.1 (1998) and BOSCHI *et al.* (2000) catalogues are compared

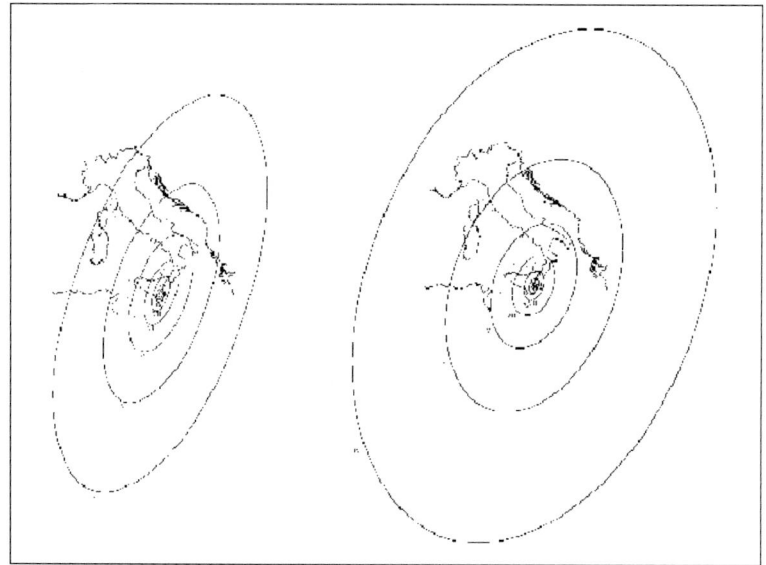

Figure 3

Felt area of the eastern Sicily 1169, February 4th earthquake (left) and eastern Sicily 1693, January 11th one (right).

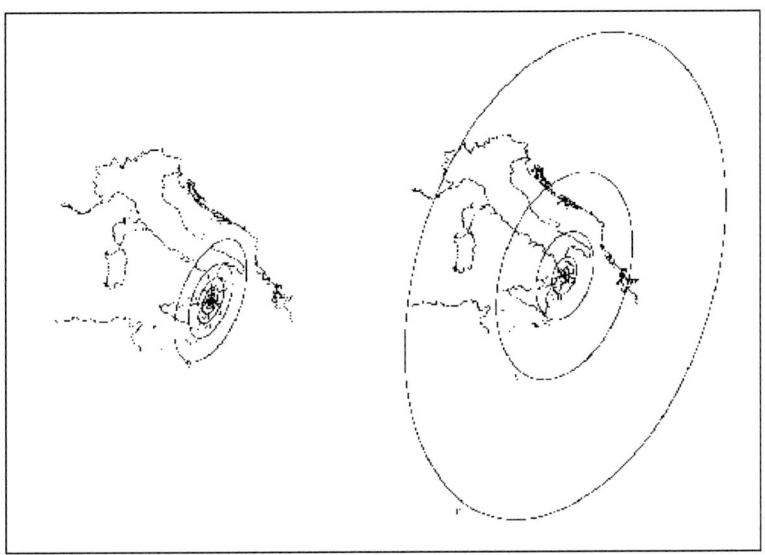

Figure 4
Felt area of the southern Calabria 1783, February 5th earthquake (on left) and central Calabria 1783, March 28th one.

with magnitude calculated with the proposed relation. For the 4-2-1169 and 11-01-1693 earthquakes, which occurred in southern Sicily, the magnitude values indicated in the NT4.1.1 catalogue are equal to 7.3 and 7.0, respectively, whereas the ones indicated in the BOSCHI *et al.* (2000) catalogue are equal to 6.5 and 7.4. The magnitude values determined with the proposed relation are equal to 7.3 and 7.4 and are also consistent with the respective extent of felt areas, equal to 39×10^5 km^2 and 42×10^5 km^2 (Fig. 3). However it is possible to observe from diagrams of Figure 2, that the magnitude of the 1169 earthquake is underestimated by the Boschi catalogue.

Therefore, for the earthquakes which occurred in southern Calabria (05-02-1783) and in central Calabria (28-03-1783) and for the 28-12-1908 Messina earthquake, the BOSCHI *et al.* (2000) catalogue shows the magnitude values equal to 6.9, 6.9 and 7.1, respectively, whereas the N.T.4.1.1 catalogue shows magnitude values equal to 7.3, 6.7 and 7.1, respectively. The magnitude calculated with the proposed relation is instead equal to 6.3, 7.4 and 6.8 consistent with the extent of the corresponding felt areas, equal to 49×10^5 km^2, 2.5×10^5 km^2 and 8.4×10^5 km^2 (Fig. 4), respectively. However it is possible to observe from diagrams of Figure 2, that the magnitude values of the 5-2-1783 and 28-3-1783 earthquakes in both catalogues are overestimated and underestimated, respectively. Instead, the values of the 1908 Messina earthquake are still overestimated in both catalogues.

It is to be observed at least for this purpose, that the differences noticed between magnitude values indicated in the catalogues considered and those determined with the proposed relation, are greater for historical earthquakes with a high epicentral intensity (XI MCS), owing to a reduced and inhomogeneous distribution of observed intensity data not directly related to the corresponding felt area, as, on the contrary, proposed here through local virtual intensity distribution. Such differences decrease in the case of recent earthquakes, owing to a greater number of observed intensity available data.

The results achieved through the proposed relation, depicted on a set of felt areas and corresponding intensity values, highlight the limits of magnitude evaluations carried out with the Sibol relation and through epicentral intensity only and, at the same time, the reliability of the proposed procedure. In fact, in the case of historical seismic events with high epicentral intensity, such as the XI grade ones which occurred in Calabria and Sicily and analyzed above, different magnitude values have been calculated with respect to the extent of felt areas, consistent with the structural framework of the respective seismogenic zones and the local geomorphological conditions. In particular, for the 1169 February 4th earthquake which occurred in eastern Sicily, whose magnitude, estimated with the Sibol relation in 5.6 and 6.5 in the two recent versions of the 1997 and 2000 Boschi catalogues (BOSCHI *et al.*, 1997, 2000), respectively, is actually still underestimated due to high epicentral intensity (XI MCS) and the lack of a felt area related to tectonic and geomorphological characteristics of the zone in which it falls, actually adopted with the proposed relation.

4. Conclusions

The relation proposed for the determination of magnitude of historical earthquakes has been related to their felt area obtained through the local virtual intensity distribution, deduced with vectorial procedures from observed intensity data and reshaped through different types of filtering. Such an approach allowed a specific magnitude evaluation for I_0 epicentral intensity levels of the earthquake in study that fall within the interval $IV \leq I_0 \leq XI$ MCS, even in cases of a reduced number of observed intensity data, actually overcoming the typical limits of procedures that use the felt area for the magnitude evaluations.

The results achieved have highlighted the applicability of the proposed procedure, with reference to both the comparison of the felt area extent of each earthquake with that of earthquakes with close epicentral intensity, however, above all, with reference to the possibility of correlating the felt area to the tectonic and geomorphological characteristics of the seismogenic zone of the earthquake in study.

REFERENCES

BOSCHI, E., GUIDOBONI, E., FERRARI, G., VALENSISE, G., and GASPERINI, P. (1997), *Catalogo dei forti terremoti in Italia 461 a.C. – 1997*, Istituto Nazionale di Geofisica. SGA.

BOSCHI, E., GUIDOBONI, E., FERRARI, G., MARIOTTI, D., VALENSISE, G., and GASPERINI, P., (2000), *Catalogne of Strong Italian Earthquakes from 461 B.C. to 1997*, Annali di Geofisica *3*, (4).

CAMASSI, R. and STUCCHI, M. (eds.) (1998), *NT4.1 un catalogo parametrico di terremoti di area italiana al di sopra della soglia di danno*, *GNDT*, Internal Report.

NUTTLI, O.W., and ZOLLWEG, J.E. (1974), *The Relation between the Felt Area and Magnitude for Central United States Earthquakes*, Bull. Seismol. Soc. Am. *69*, 893–909.

SIBOL, M.S., BOLLINGER, G.A., and BIRCH, J.B. (1987), *Estimation of Magnitudes in Central and Eastern North America Using Intensity and Felt Area*, Bull. Seismol. Soc. Am. *77*, (5), 1635–1654.

TERAMO, A., STILLITANI, E., and BOTTARI, A. (1995), *Anisotropic Characterization of Macroseismic Fields*, Natural Hazards, *11*, 223–245.

TERAMO, A., TERMINI, D., STILLITANI, E., and BOTTARI, A. (1996), *The Determination of the Epicentre by a Vectorial Modelling of Macroseismic Intensity Distribution*, Natural Hazards, *13*, 101–117.

TERAMO, A., TERMINI, D., STILLITANI, E., and BOTTARI, A. (1998), *An Anisotropic Modelling for the Determination of the Regional Attenuation Coefficients*, Natural Hazards *17*, 17–30.

TERMINI, D., BOTTARI, A., TERAMO, A., and TUVÈ, T. (2005a), *On the Observed Intensity Filtering in the Anisotropic Modelling of Macroseismic Intensity*, Pure Appl. Geophys. *162*, 683–697.

TERMINI, D., TERAMO, A., BOTTARI, C., and TUVÈ, T. (2005b), *An Anisotropic Attenuation Law of Macroseismic Intensity Performed on Virtual Intensity Distribution of Seismogenic Zones*, Pure and Appl. Geophys. *162*, 707–714.

TERMINI, D., TERAMO, A., and BOTTARI, C. (2005c), *On the Intensity Virtual Area Characterization in the Intensity Distribution Modelling by Macroseismic Planes*, Pure and Appl. Geophys. *162*, 699–705.

(Received May 18, 2001, revised October 4, 2001, accepted January 16, 2002)

 To access this journal online:
http://www.birkhauser.ch

Pure appl. geophys. 162 (2005) 739–746
0033–4553/05/040739–8
DOI 10.1007/s00024-004-2637-8

© Birkhäuser Verlag, Basel, 2005

❘Pure and Applied Geophysics

Macroseismic Parameters of Seismogenic Zones of Calabria and Sicily for Seismic Hazard Evaluation

CARLA BOTTARI[1], DOMENICA TERMINI[1], and ANTONIO TERAMO[1]

Abstract—For the seismic hazard evaluation of the region including the southern Calabro-Peloritanian Arc and southeastern Sicily, the determination of the macroseismic virtual intensity distributions has been carried out, characteristic of the seismogenic zones that fall within the area in study, starting with the structural framework of the region and from the analysis of the observed intensity effected through suitable filters. The macroseismic parameters, derived from such virtual distributions and used for seismic hazard evaluation, are not only a reference for eventual subsequently deeper knowledge referable to the need for a better characterization of the reference modelling, but distinguish themselves as an essential instrument for the definition of seismic hazard scenarios correlated to seismic events that take place in single seismogenic zones.

Key words: Seismogenic zones, virtual intensity distribution, seismic hazard.

1. Introduction

For many years the geodynamics of southern Italy and, in particular, the Calabro-Peloritanian Arc one, have been of great scientific interest. Researchers of Earth Sciences have found a natural laboratory here, suitable for studying the tectonic processes and related phenomenologies. Peculiar features of this seismic region are the high maximum energy, (magnitude up to 7.2) (SCHICK, 1977; CAMASSI and STUCCHI, 1997), and the quantity and variety of effects produced on the ground surface.

Fissures, landslides, soil liquefaction and collapse phenomena accompanied the main earthquakes (MERCALLI, 1897; BARATTA, 1909, 1910; LO PERFIDO, 1909) and some of these events were also followed by tsunamis (BARATTA, 1910; CAPUTO and FAIFA, 1984). Furthermore, the size and spatial distribution of the surface phenomena area are indicative of the shallowness of the seismic sources, and in agreement with the earthquake focal depth estimated by instrumental and macro-seimic data (BOTTARI et al., 1979).

[1] Osservatorio Sismologico, Università di Messina,Via Osservatorio, 4 – 98121 Messina, Italy.
E-mail: cbottari@unime.it

The scientific interest for the seismicity concerning this territory has been enlarged by consequent problems with the associated high risk level. The results proposed in this paper derive from recent in-depth studies on the analysis of observed intensity, carried out through 1st and 2nd level filtered macroseismic planes (TERMINI *et al.*, 2005a), that allow seismic hazard evaluations correlated to a greater reliability of the macroseismic parameters of input.

2. Geostructural Remarks

As is reported in the catalogues of Italian earthquakes (POSTPISCHL, 1985; CAMASSI and STUCCHI, 1997), the strongest earthquakes over the last few centuries concerning the Italian territory occurred in the region bounded to the north by the Gulf of S.Eufemia (Calabria) and to the south by the Nebrodi Chain (northeastern Sicily) (Fig. 1).

This area is geographically placed in one of the sectors more subjected to an affine strain of southern Italy. Here the collision of the African Plate and the European plate have determined important strain processes since the early Miocene age.

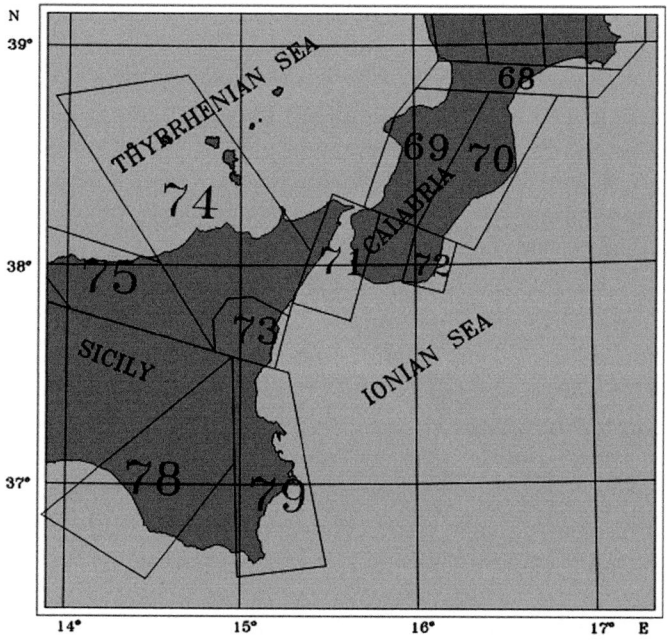

Figure 1
Seismogenic zonation of southern Italy.

In the Mediterranean Sea basin, between the Jurassic and middle Cretaceous eras, a radical change of stress levels occurred, and the African margin from being expansive became compressive. This determined the Tetide closure that caused the phase of subduction of the African margin below the European margin. During the late Miocene age, coinciding with the opening of the western Mediterranean and Tyrrhenian basin's, an anticlockwise rotation of the Italian peninsula took place and produced strong distortions in the Calabro-Peloritanian arc with compression in the central inner zone and lateral stretching on the borders.

The post-Orogenic tectonic phases beginning in the Upper Miocene era and particularly the more markedly distensive ones of the Upper Pliocene era onward, divided the chain into transverse and longitudinal blocks. The longitudinal faulting gave an antiform feature to the Calabrian-Peloritanian Arc, with axial grabens in the hinge zone; the transverse faulting is characterized by fan-shaped grabens which separate the blocks, giving rise to the arch-shaped structure.

In eastern Sicily several structures of regional importance developed (LENTINI et al., 1995) belonging to different systems: the *"Sistema Sud-Tirrenico"* with NW-SE and WNW-ESE orientation, Plio-pleistocenic aged, represented by compound dextral faults (LENTINI et al.,1996), the NNW-SSE system that delimits the Ionian margin of eastern Sicily and forms the Ibleo-Maltese escarpment, to which the opening of the Ionian basin is attributed, and that had a normal reopening during the Plio-Pleistocene age (CARBONE et al., 1982; LENTINI et al., 1995); the Messina-Etna system, with prevalently normal faults, is located in accordance with a NNE-SSW direction (LENTINI et al., 1996). This last system close to the Messina Straits appears as a complex strike zone, characterized by active structures in the middle Pliocene age whose geometry suggests a dextral component of movement along the principal strike (LENTINI et al., 1996).

Over the last few years geological studies and macroseismic data analyses have enlarged knowledge of the seismotectonic characteristics of this region. Among the most studied problems is the modelling of macroseismic intensity aimed at seismic hazard evaluation.

3. Macroseismic Intensity Modelling

The modelling of macroseismic intensity, aimed at seismic hazard evaluation of a given region, in accordance with an approach recently proposed (HOLSCHNEIDER et al., 2004), must be carried out through a preventive reconfiguration of observed intensity data by procedures that allow the characterization of intensity distributions that are, on one hand, consistent with the observed data, geostructural framework of the area in study and with an anisotropic attenuation law and, on the other, representative of the seismicity of the single seismogenic zones.

Such procedures are essentially linked to the determination of the 1st and 2nd level filtered macroseismic planes (TERMINI et al., 2005a, b) of earthquakes under study, to the local and regional virtual intensity distributions (TERMINI *et al.*, 2005c; TERAMO *et al.*, 1998) and to the application of an attenuation law shaped on the characteristics of seismogenic zones (TERMINI *et al.*, 2005c).

With greater detail, starting with the observed intensity data, the construction of macroseismic planes of earthquakes belonging to seismogenic zones that individualize the region in study, is carried out. Such macroseismic planes are subsequently reshaped through procedures of approximation of half degrees and integer ones, to allow for eventual evaluation errors of the observed intensity, depicting, respectively, 1st and 2nd level filtered macroseismic planes. The epicenters and attenuation directions of single seismic events are therefore brought about through a vectorial procedure (TERAMO *et al.*, 1995a,b 1996), verifying the consistency respectively with the maximum shaking areas and the corresponding structural lineaments.

The definition of the intensity distributions of the events placed in a given seismogenic zone, subjected moreover to further single verifications through the respective *local* virtual intensity distributions (TERMINI *et al.*, 2005a), allows the determination of the *regional* virtual intensity distribution (TERAMO *et al.*, 1998) for seismogenic zones that, being characteristic for each seismogenic zone, represents the reference for seismic hazard evaluations. From such an intensity distribution, characterized by a number of similar ellipses in which center coincides with the virtual epicenter of the seismogenic zone and semi-axes oriented towards minimum and maximum attenuation directions (characteristic of the seismogenic zone the parameters for seismic hazard evaluations of the region in study are established.

4. *Data and Analyses*

The modelling of the intensity distribution has been carried out on a region concerning seismogenic zones individualized by the numbers 69, 70, 71, 72, 74, 75, 78 and 79 in the seismic zonation (SCANDONE *et al.*, 1992) adopted from GNDT (Gruppo Nazionale per la Difesa dai Terremoti, C.N.R., Rome) for the seismic hazard of Italy in the last version of April 1996 (Fig. 1).

The earthquakes analyzed for the determination of macroseismic parameters of seismogenic zones occurred between the 12th and 20th centuries, with epicentral intensities $I_0 \geq V$ MCS and extensions relatively wide of the shaking area and a suitable number of observed intensity points, are reported in Table 1 in which the geographic epicentral coordinates, time origin, maximum intensity, geographic site of reference, and the references are listed.

The macroseismic parameters calculated for each considered seismogenic zone considered, reported in Table 2, are:

$$\rho_{o\max}, \rho_{o\min}, K, \theta,$$

where: $\rho_{o\max}$ and $\rho_{o\min}$ are the largest and the smallest semi-axes of the first virtual intensity ellipse; K, the amplification factor of ellipses through which it is possible to

Table 1

List of earthquakes analysed for the determination of the macroseismic parameters

Ye	Mo	Da	Ho	Mi	Site	I_o	Lat.	Lon.	Sz	References
1783	3	28			CALABRIA CENTR.	100	38.800	16.467	68	Boschi et al., 1997
1659	11	5			LE SERRE	95	38.700	16.333	69	Boschi et al., 1997
1783	2	5			CALABRIA MERID.	110	38.267	15.917	69	Boschi et al., 1997
1783	2	7			SORIANO SERRE	105	38.583	16.217	69	Boschi et al., 1997
1791	10	13	1	20	LE SERRE	90	38.659	16.248	69	Boschi et al., 1997
1869	11	28			VIBO VALENTIA	65	38.700	16.117	69	Boschi et al., 1997
1889	10	5	13	52	TROPEA	55	38.683	15.900	69	Boschi et al., 1997
1894	11	16	17	52	BAGNARA CALABRA	85	38.278	15.903	69	Boschi et al., 1997
1905	9	8	1	43	GOLFO DI S.EUFEMIA	110	38.754	16.026	69	Boschi et al., 1997
1947	5	11	6	32	MAR IONIO	85	38.712	16.581	70	Boschi et al., 1997
1908	12	28	4	20	CALABRO MESSINESE	110	38.133	15.667	71	Bottari et al., 1986
1975	1	16	0	9	REGGIO CALABRIA	70	38.117	15.650	71	Boschi et al., 1997
1907	10	23	20	28	FERRUZZANO	85	38.155	16.024	72	Boschi et al., 1997
1978	3	11	19	20	BRANCALEONE		37.967	16.183	72	Boschi et al., 1997
1786	3	9			PATTI	85	38.100	15.000	74	Boschi et al., 1997
1908	12	10	6	20	NOVARA DI SICILIA	70	38.017	15.083	74	Boschi et al., 1997
1978	4	15	23	33	GOLFO DI PATTI	85	38.150	14.983	74	Boschi et al., 1997
1818	9	8			MADONIE	75	37.833	14.083	75	Boschi et al., 1997
1967	10	31	21	8	SPERLINGA	80	37.850	14.367	75	Boschi et al., 1997
1968	1	15	2	1	VALLE DEL BELICE	95	37.750	12.967	77	Boschi et al., 1997
1818	2	28	8	30	MINEO	75	37.250	14.700	78	Boschi et al., 1997
1892	1	23	0	46	LICODIA	55	37.150	14.700	78	Boschi et al., 1997
1169	2	4	7	0	SICILIA ORIENTALE	110	37.333	15.200	79	Boschi et al., 1997
1542	12	10	16	0	SORTINO	95	37.250	15.067	79	Boschi et al., 1997
1693	1	11	8	0	SICILIA ORIENTALE	75	37.330	14.996	79	Barbano and Cosentino, 1981
1990	12	13	00	24	SICILIA SUD-ORIENTALE	75	37.270	15.070	79	Boschi et al., 1997

Table 2

Macroseismic parameters for seismogenic zones

Zone	ρ_{max}	ρ_{min}	θ	I_{max}	k
68	2,54	1,48	287	11	5,23
69	7,91	3,64	335	11	2,12
70	9,16	4,39	343	8	5,35
71	14,97	10,25	337	11	2,58
72	8,31	4,24	333	8	4,01
74	11,85	5,74	25	9	2,36
75	5,79	3,75	36	8	3,83
77	4,53	1,71	22	10	4,07
78	19,13	11,7	325	7	3,84
79	8,92	5,82	332	11	2,14

arrange all the ellipses of the virtual intensity distribution with reference to the ratios $\rho_{i\ max} / \rho_{0\ max} = \rho_{i\ min} / \rho_{0\ min} = K^{i/2}$; θ the angle between $\rho_{0,min}$ (that singles out the maximum attenuation direction) and the x axis of the system of reference $C(x, y)$ (Fig. 2).

The calculation of such parameters, necessary for seismic hazard evaluations (HOLSCHNEIDER *et al.*, 2004), is referable to a modelling of observed intensity produced through 1st and 2nd level filtered macroseismic planes, and it allows a more reliable characterization of the virtual intensity distribution and the consequent definition of the attenuation typology.

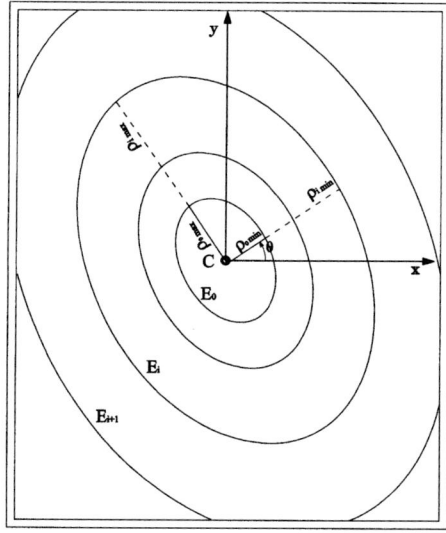

Figure 2
Geometric representation of a virtual intensity distribution.

5. Conclusions

The definition of the macroseismic parameters for the characterization of virtual intensity distribution, characteristic of the seismogenic zone of the region in study, represents the requirement for seismic hazard evaluations.

Recent developments relative to the observed intensity modelling, referable to the application of suitable filters on half degrees and integer ones, have produced a more accurate calculation of the macroseismic parameters derived through a configuration of the virtual intensity distribution. Such parameters, so deduced, represent the reference for eventual following in-depth studies correlated to the need for a better characterization of the seismic hazard scenarios of areas belonging to single seismogenic zones, referred to as the occurrence of events originating in the same areas.

The pursuit of such an aim now is not only consistent with the requirements of seismological research, but is an instrument with strong applicable value for the predisposition of suitable civil protection interventions. The delay and the inadequacy of first aid, in consequence of the lack of correct macroseismic modelling with the backing of the definition of expected damage scenarios, have been tragically highlighted in Italy in the occurrence of the strongest earthquakes in the second half of the 20th century. Particularly serious was the situation after the Irpinia earthquake (southern Italy, 20-11-1980) resulting from the inadequacy of intervention and due to the lack of a preventive organization of a road system of the same on the basis of suitable knowledge of seismic intensity distribution on more damaged territories.

REFERENCES

BARATTA, M. (1909), Il terremoto calabro-siculo del 28 Dicembre 1908, Boll. Soc. Geogr. It., S. IV, X, (8), 852–882; (9), 990–1010.

BARATTA, M. (1910), La catastrofe sismica calabro-messinese (28 dicembre 1908), Relazione alla Soc. Geogr. It., Roma

BARBANO, M.S. and COSENTINO, M. (1981), Il terremoto siciliano dell'11-1-1693, Rend. Soc. Geol. It. 4, 517–562

BOSCHI, E., GUIDOBONI, E., FERRARI, G., VALENSISE, G., and GASPERINI, P. (1997), Catalogo dei forti terremoti in Italia dal 461 a.C. al 1990. I.N.G., S.G.A. Storia Geofisica ambiente, 2.

BOTTARI, A., FEDERICO, B., and LO GIUDICE, E. (1979), Methodological Consideration Regarding the Determination of Some Macroseismic Field Parameters. Application to Earthquakes in the Calabro-Peloritan Arc, Boll. Geof. Teor. Appl. XXI (83), 197–225.

BOTTARI, A., CARAPEZZA, E, CARAPEZZA, M., CARVENI, P., CEFALI, F., LO GIUDICE, E., and PANDOLFO, C. (1986), The 1908 Messina Strait Earthquake in the Geostructural Framework, J. Geodyn. 5, 275–302.

CAMASSI, R., and STUCCHI, M. (eds.) (1997), NT4.1 un catalogo parametrico di terremoti di area italiana al di sopra della soglia di danno. GNDT, Internal Report, 84 pp.

CAPUTO, M. and FAITA, G. (1984), Primo catalogo dei maremoti delle coste italiane, Atti Acc. Naz. Lincei. Mem. Classe di Sc. Fis., Mat. E Nat., S. VIII, 17, Sez. 1, 213–356

CARBONE, S., GRASSO, M., and LENTINI, F. (1982), Considerazioni sull'evoluzione geodinamica della Sicilia sud-orientale dal Cretaceo al Quaternario. Mem. Soc. It. 24, 362–386.

HOLSCHNEIDER, M., TERAMO, A., BOTTARI, A., and TERMINI, D. (2004), *On the Relevance of the Spatial Distribution of Events for Seismic Hazard Evaluation,* Natural Hazards, *31,* 1–19.

LENTINI, F., CARBONE, S., CATALANO, S., DI STEFANO, A., GARBANO, C., ROMEO, M., STRABUZZA, S., and VINCI, G. (1995), *Sedimentary Evolution of Basin in Mobile Belts: Examples from Tertiary Terrigenous Sequences of Peloritani Mts. (NE Sicily),* Terra Nova 7, (2), 10.

LENTINI, F., CARBONE, S., CATALANO, S., and GRASSO, M. (1996), *Elementi per la ricostruzione del quadro strutturale della Sicilia Oriental,.* Mem. Soc. Geol. It., *51,* 179–195.

LO PERFIDO, A. (1909), *Livellazione geometrica di precisione eseguito dall'istituto geografico nazionale sulla costa orientale della Sicilia, da Messina a Castanea, a Gesso ed a Faro Peloro e sulla costa occidentale della Calabria da gioia Tauro a Melito di Porto Salvo, per incarico del Ministro dell'agricultura, industria e commercio,* Relaz. Comm. R., 131–169.

MERCALLI, A. (1897), *I terremoti della Calabria meridionale e del messinese,* Mem. Soc. It. Sc. Ser. III, *XL,* R . Acc. Lincei, Roma.

POSTPISCHL, D. (1985), Catalogo dei terremoti italiani, dall'anno 1000 al 1980. CNR, P.F. Geodinamica Gratico Bologna, 239 pp.

SCANDONE, P., PATACCA, E., MELETTI, C., BELLATALLA, M., PERILLI, N., and SANTINI, U. (1992), *Struttura Geologica, evoluzione cinematica e schema sismotettonico della Penisola Italiana,* Atti del convegno GNDT, Pisa 25–27 Giugno 1990,*1,* 119–135. Edizione Ambiente, Bologna.

SCHICK, R.(1977), *Eine seismotektonische Bearbeitung des Erdbebens van Messina im Jahre 1908,* Geol. Jahrb., *E 11,* 3–74.

TERAMO, A., STILLITANI, E., and BOTTARI, A. (1995a), *Anisotropic Characterization of Macroseismic Fields,* Natural Hazards *11,* 223–245.

TERAMO, A., STILLITANI, E., and BOTTARI, A. (1995b), *An Anisotropic Attenuaion Law of Macroseismic Intensity,* Natural Hazards *11,* 203–221.

TERAMO, A., TERMINI, D., STILLITANI, E., and BOTTARI, A. (1996), *The Determination of Epicentre by a Vectorial Modelling of Macroseismic Intensity Distribution,* Natural Hazards *13,* 101–117.

TERAMO, A., TERMINI, D., STILLITANI, E., and BOTTARI, A. (1998), *An Anisotropic Modelling for the Determinaion of Regional Attenuation Coefficents,* Natural Hazards *17,* 17–30.

TERMINI, D., TERAMO, A., TUVE, T., and BOTTARI, A. (2005a), *On the Observed Intensity Filtering in the Anisotropic Distribution Modelling of Macroseismic Intensity,* Pure and Appl. Geophys. *162,* 683–697.

TERMINI, D., TERAMO, A., and BOTTARI, C. (2005b), *On the Intensity Virtual Area Characterization in the Intensity Distribution Modelling by Macroseismic Planes,* Pure and Appl. Geophys. *162,* 699–706.

TERMINI, D., TERAMO, A., BOTTARI, C., and TUVÈ, T. (2005c), *An Anisotropic Attenuation Law of Macroseismic Intensity Performed on Virtual Intensity Distribution of Seismogenic Zones,* Pure and Appl. Geophys. *162,* 707–714.

(Received July 27, 2001, revised October 18, 2001, accepted December 26, 2001)

To access this journal online:
http://www.birkhauser.ch

Pure appl. geophys. 162 (2005) 747–760
0033–4553/05/040747–14
DOI 10.1007/s00024-004-2638-7

© Birkhäuser Verlag, Basel, 2005

| Pure and Applied Geophysics

Macroseismic Intensity Modelling of Earthquakes in Southern Spain: Attenuation and Tectonic Implications

TIZIANA TUVÈ[1], ANTONIO TERAMO[1], JESUS IBAÑEZ[2], CARLA BOTTARI[2],
and DOMENICA TERMINI[1]

Abstract—A modelling of the observed macroseismic intensity of historical and instrumental earthquakes in southern Spain is proposed, with the aim of determining the macroseismic parameters for seismic hazard evaluation in a region in which the characterization of intensity distribution of seismic events shows different levels of difficulty referable to the complex faults system of the area in study.

The adopted procedure allows an analytical determination of epicenters and principal attenuation directions of earthquakes with a double level of verification with reference to the maximum shaking area and structural lineaments of the region, respectively. The analyses, carried out on a suitable number of events, highlight, therefore, some elements for a preliminary characterization of a seismic zonation on the basis of the consistency between seismic intensity distribution of earthquakes and corresponding structural framework.

Key words: Attenuation directions, southern Spain, macroseismic intensity, virtual intensity.

1. Introduction

The aim of this paper is the characterization of seismicity of southern Spain through a methodological approach which, starting with the intensity maps of earthquakes belonging to the region in study, models the intensity distribution with several procedures that join an efficient reconfiguration of the observed intensity to a verification of its consistency with the areas of maximum shaking and structural lineaments.

The adoption of this approach, already tested with good results in the intensity modelling of several seismogenic zones of southern Italy and Sicily, is referable to the complex tectonic system of the region in study, to the lack of a seismic zonation of territory, and to the particular distribution of intensities of earthquakes; some of which are close to the coast and have a reduced number of intensity points that do not allow an easy definition of epicenters and principal attenuation directions.

[1] Osservatorio Sismologico, Università di Messina, Via Osservatorio, 4 - 98121 Messina, Italy.
[2] Instituto Andaluz de Geofisica, Universidad de Granada, Campus de Cartuja s/n 18071, Granada, Spain
Corresponding author: Tuvè T., e-mail: tuvet@unime.it

In this ambit it has to be highlighted that 18 strong earthquakes with a magnitude greater than 6.0 (IBAÑEZ *et al.*, in preparation) have occurred since the 9th century, and many intensity attenuation laws for the entire Iberian Peninsula have been obtained (LOPEZ-CASADO *et al.*, in preparation), showing a complex attenuation pattern of the area with different behavior (from very high to very low attenuation).

In particular, the procedure, based on a modelling of intensity distribution called macroseismic plane (TERAMO *et al.*, 1996) is different from the macroseismic field plane, and consists of several areolas build around the points of the intensity map. It allows specific evaluations of the reliability of intensity associated with the points of the intensity map and an approximation of half and integer degrees (TERMINI *et al.*, 2005a) using suitable filters that originate the so-called 1st and 2nd level filtered macroseismic planes. This procedure distinguishes itself because it allows an analytical determination of epicenter and principal attenuation directions whose consistency, respectively, with the area of maximum shaking and the direction of structural lineaments, can be easily verified.

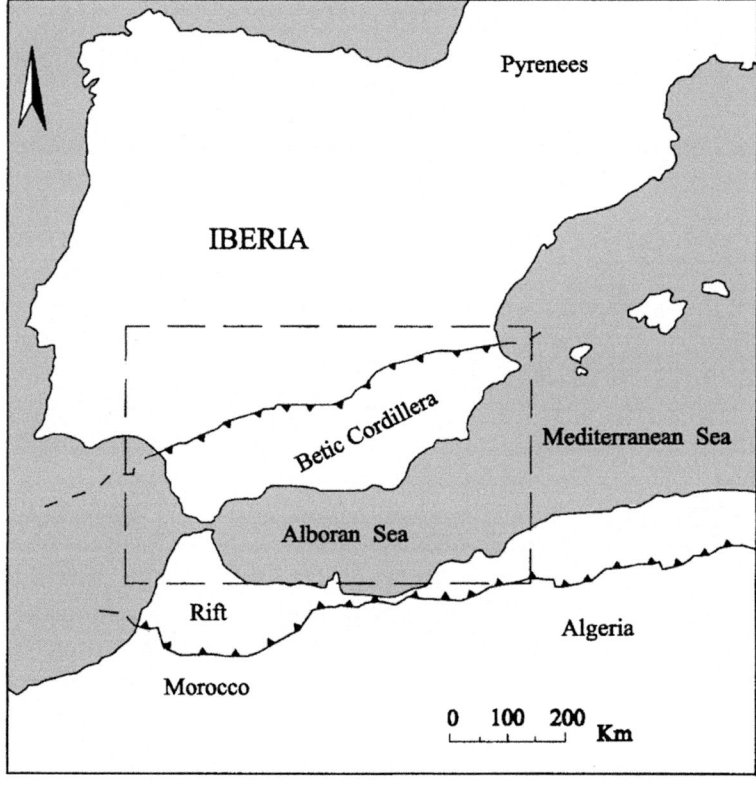

Figure 1
Seismotectonic framework of the Ibero-Maghrebian region.

A further approximation of intensity distribution could be obtained through the local virtual intensity distribution (TERMINI et al., 2005a) individualized by a set of similar ellipses, whose axes are oriented according to the directions of maximum and minimum attenuation and whose centers coincide with the epicentre of the earthquake.

As the area in study does not have a specific subdivision in seismogenic zones, and considering that the geological frame of the region in study distinguishes itself due to a remarkable complexity, the aim to characterize preliminarily a seismic zonation of territory can be reached by individualizing sets of homogeneous distributions of local virtual intensity, in terms of attenuation directions, relative to the earthquakes in study.

2. Geological Frame

Many authors have studied the tectonic and geodynamic characteristics of southern Spain (BOCCALETTI et al., 1987; SANZ DE GALDEANO et al., 1995; GONZALEZ-LODEIRO et al., 1996; LONERGAN and WHITE, 1997), nonetheless many problems have not been solved due to the complex spatial fault distribution. Spatial earthquake distribution is sparse, and a complex stress and a paleostress pattern have been observed (GALINDO-ZALDIVER et al., 1999). In such a context, some recent results obtained from several authors for southern Spain, related to tectonic and geodynamic studies, must be highlighted.

The Betic Cordillera and the Rift in Morocco constitute the Gibraltar Arc; the Betic Cordillera, furthermore, represents one of the most evident oroclins of the chain alpine system on the western Mediterranean area (Fig. 1). The tectonic evolution of the Betic Belt occurred during the long process of the opening of the Atlantic Ocean and the consequent and continuous approach of the African and European plates. In the Betic Cordillera the principal tectonic units that constitute the structuring of the chain are: Inner Zone and Outer Zone.

The Outer Zones are subdivided into the Prebetic and Subbetic and consist of Mesozoic and Tertiary formations; the Inner Zone, or the Betic Zone s.s., consists of three tectonic units: the Nevado–Filabride unit, the Alpujarride and the Malaguide unit (BOCCALETTI et al., 1987). These units were structured in part during the neo-alpine phase, and in the Burdigalian were superimposed on the Outer Zone with a dextral transpressive movement. In this phase the Subbetic zone is involved in folding and in thrust to the Prebetic Zone.

Successively, during the middle and late Miocene era, the intra-chain were influenced by important faults with a transcurrent component. In fact, in this period the genesis of the intra-chain sedimentary basins can be considered in a context of N-S regional compression. On the basis of meso- and macro-structural field data, the structural evolution of the Betic Cordillera in the Neogene-Quaternary era, was characterized by a compression with directions NNW-SSE and N-S due to the European-African plate convergence. In particular, during the Quaternary era, the N-S compression originated a conjugate system of transcurrent faults.

GALINDO-ZALDIVAR *et al.* (1999), in their study on the stress state of the area, have shown the existence at present of a heterogeneous stress state that is not consistent with the results observed in the focal mechanisms obtained in the region. These authors determined the paleo-stresses from the study of the orientation and kinematics of micro faults. They took into account the orientation of the major structures and the results derived from the study of earthquakes with a magnitude greater than 2.5, verifying that both geological surface data and focal mechanism indicate regional conditions of NE-SW extension. However, the stress field is heterogeneous with local variations over time, with different stresses sometimes simultaneously acting in adjacent areas. Moreover, there is a NE-SW extension and a NW-SE sub-horizontal compression. Strike-slip faults are scarce even though they are the most likely structures to be expected in a region with SW-NE extension and NW-SE extension.

BLANCO and SPAKMAN (1993) found a high-velocity anomaly beneath the Betic-Rift orogenic belt in a 200–650 km depth range. These authors suggested the presence of a slab, probably an old oceanic lithosphere fragment. MORALES *et al.* (1999) found a velocity anomaly beneath the Malaga area extending to 100 km in

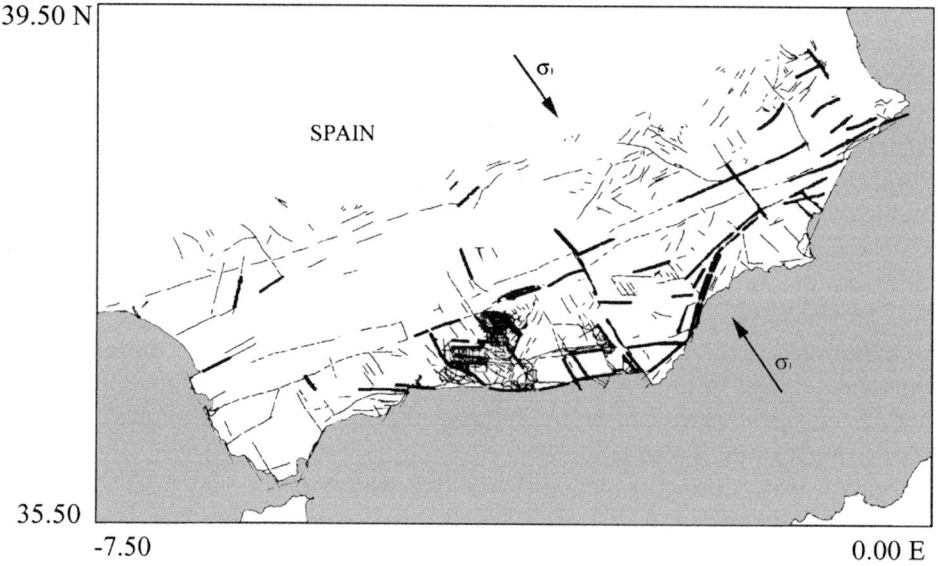

Figure 2
The Betic Cordillera with map of the fracture network (SANZ DE GALDEANO, 1995).

depth. They interpreted this anomaly as a probable continental subduction in the region, a subduction that is active at present.

In southern Spain three main transcurrent fault systems of different importance, acting until the Quaternary era, are recognized: 1) system *A*, with an orientation between E-W and N70 °, and showing important dextral movements; 2) system *B*, oriented N120 °–150 °, with prevailingly dextral movements; 3) system *C*, oriented NNE/NE with prevailingly sinistral movement (Fig. 2) (BOCCALETTI *et al.*, 1987).

System *A* is represented by the longest and most important fractures of the Betic Cordillera. Its direction marks, during the Burdigalian, with transpressive regime, the contact between the Inner and Outer Zone. Also the contact Inner/Outer Zones corresponds to one of these lines, which produced an E-W translation of at least 300 km of the Inner Zone. Other lines of this system, e.g., the Alicante-Cadiz and the Alpujarran Corredor, had a similar geodynamic role.

The genesis of the Neogene intra-chain sedimentary basins, such as the Guadalquivir depression, in the external position (Foreland Basin) and adjacent to the Meseta, the basins of Granada, Guadix, Baza and Ronda, situated between the Inner and Outer Zone, the basins of Almeria, Sorbas-Vera and others in the eastern Inner Zones, can be interpreted in the context of the N-S regional compression. The genesis of the Alboran Sea could also be included in this context. It can represent a complex system of pull-apart basins mainly related to the *A* and *C* systems (BOCCALETTI *et al.*, 1987).

Since the Miocene era, both the Alboran Sea and the Cordilleras have undergone extensional tectonic activity that has produced a thinning of the thickened continental crust (e.g., GONZÁLEZ-LODEIRO *et al.*, 1996). Compressional structures can also be observed in the area: folds and thrust in the external region of the cordilleras and folds and strike-slip faults in the internal zones. In recent years several geodynamic models have been proposed to explain continental extension in the area within a general compressional context: rollback processes (e.g., LONERGAN and WHITE, 1997), mantle diapirism (e.g., WEIJERMARS, 1985), extensional collapse (e.g., VISSERS *et al.*, 1995), delamination (e.g., SEBER *et al.*, 1996) or continental subduction (e.g., MORALES *et al.*, 1999).

We can consider the sedimentary tectonic evolution of the Betic Cordillera in the context of the Gibraltar Arc, as an effect of the lateral extrusion toward WSW as a consequence of the advance of the African plate towards the European plate (BOCCALETTI *et al.*, 1987). The structural complexity of the chain gives a problematic seismotectonic interpretation of the seismicity and, for this reason it is necessary to adopt a suitable intensity distribution modelling, as a preliminary approach to seismic hazard evaluation.

3. The Macroseismic Intensity Distribution Modelling

3.1. Historical Earthquakes

The element of greater evidence that arises from an analysis of historical data is referable to the incompleteness of the catalogue of historical earthquakes of the southern Spain region. This derives essentially from fragmenting, non homogeneous and incomplete information, in particular in distant centuries, and it is strictly connected with the social and cultural context that characterized southern Spain in different historical times.

In particular, the earthquakes which occurred before the 9th century, are insufficiently substantiated and the relative documents are not reliable. For the seismic events that occurred between the 9th and the 15th century, only some descriptions of their destructive effects are available. Starting with the 15th century, the descriptions of the damage distribution that were already sufficient to determine the epicenter location (SANCHEZ NAVARRO-NEUMANN, 1921) became gradually more detailed and reliable until the 19th century.

Only recently has a complete catalogue of the strongest earthquakes that occurred in southern Spain from the 9th to 19th century (IBAÑEZ *et al.*, in preparation), with a magnitude range between 5.2 and 7.1, become available. In this catalogue, from a set of 40 earthquakes with a maximum intensity greater than VIII (MSK scale), only 6 earthquakes from 1504 to 1884, with characteristics suitable to the application of the above-mentioned procedure aimed at the macroseismic intensity modelling, have been individualized (Fig. 3).

3.2. Recorded Earthquakes

The recording of earthquakes which have occurred in southern Spain starts in 1903 with the deployment of the Cartuja station in the city of Granada. In 1980 the seismic station distribution of the region in study was enough to provide a reliable location

Table 1

List of historical earthquakes

Earthquake number	Date (y-m-d)	I_{max} (MSK)	Calculated epicenters (Lat, Long.)	Observed epicenters (Lat, Long.)	Zone
1	1504-05-04	IX	37,55; −5,7	37,24; −5,36	Carmona
2	1518-09-11	IX	37,2; −1,85	37,2; −1,9	Vera
3	1680-09-10	IX	36,75; −4,6	36,7; −4,4	Malaga
4	1804-13-01	VIII	36,7; −3,55	36,7; −3,5	Motril
5	1804-25-08	IX	36,85; −2,9	36,8; −2,8	Dalias
6	1884-25-12	X	37,05; −4,75	36,9; −4,0	Arenas del Rey

Table 2

List of recorded earthquakes

Earthquake number	Date (y-m-d)	I$_{max.}$ (MSK)	Calculated epicenters (Lat, Long.)	Observed epicentres (Lat, Long.)	Zone
7	1910-06-16	VIII	36,9; −3,2	36,7; −3,1	Adra (AL)
8	1926-11-03	VI-VII	37,0; −2,6	37,1; −2,4	–
9	1930-05-07	VIII	37,58; −4,7	37,6; −4,7	Montilla
10	1932-05-03	VIII	37,3; −2,45	37,4; −2,4	Lucar
11	1951-10-03	VIII	38,06; −4,0	38,1; −3,8	Bailèn
12	1951-19-05	VIII	37,6; −4,0	37,6; −3,9	Alcaudete
13	1954-08-01	VII-VIII	36,9; −3,85	36,9; −3,9	Arenas R.
14	1956-19-04	VIII	37,16; −3,63	37,2; −3,7	Albolote
15	1979-20-03	VI	37,25; −3,75	37,17; −3,80	Santa Fè
16	1979-20-06	VI	37,2; −3,57	37,24; −3,49	Santa Fè
17	1979-30-07	VI	37,15; −3,60	37,12; −3,65	Armilla
18	1979-25-11	VI	37,2; −3,75	36,87; −3,77	Jayena
19	1984-24-06	VII	36,96; −3,75	36,83; −3,74	—
20	1993-23-12	VII	36,5; −3,0	36,7; −3,1	Berja
21	1994-04-01	VI-VII	36,52; −2,05	36,6; −2,8	Costa Balerma

of earthquake epicenters, however the corresponding intensity maps (MEZCUA, 1982) reveal an incomplete distribution of observed intensity points due to lack of information for some villages or towns.

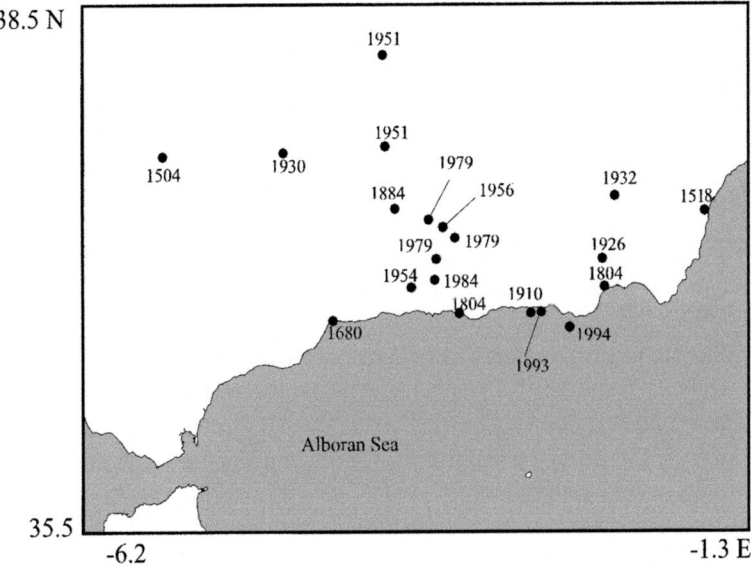

Figure 3
Map of historical and recorded earthquakes in southern Spain.

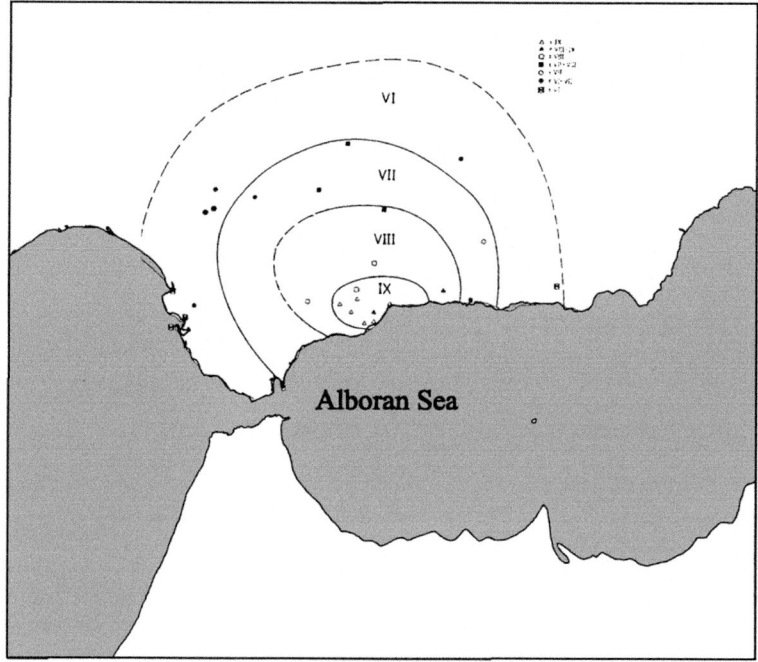

Figure 4
Macroseismic field of the 10-09-1680 earthquake (Mezcua, 1982).

Particularly, in the Mezcua catalogue of earthquakes, the intensity maps of about 15 earthquakes which occurred in this region from 1910 to 1994 have been selected for the intensity modelling, from a set of 60 earthquakes with an intensity greater then V of the MSK scale and a magnitude range between 4.0 and 6.3.

3.3. The Macroseismic Intensity Modelling

The essential characteristics of the macroseismic intensity modelling, depicted on all the seismic events reported in Tables 1 and 2, through different levels of approximation of observed intensity and subsequent reconfigurations of the corresponding macroseismic planes, for reasons of brevity, are deduced here with reference only to the 10-09-1680 earthquake (IX MSK) that took place in southern Spain (Fig. 3).

The corresponding macroseismic plane (Fig. 4), built through the intensity observed points, has been reconfigured approximating the half and integer degrees not consistent with an anisotropic model of intensity attenuation (TERMINI *et al.*, 2005a) through the application of suitable filters that originated the 1st and 2nd level filtered macroseismic planes (Fig. 5) (TERMINI *et al.*, 2005b). A further reconfiguration, linked to the inhomogeneous distribution of the observed intensity points, correlated to the location along the coast of the epicenter, is created through the

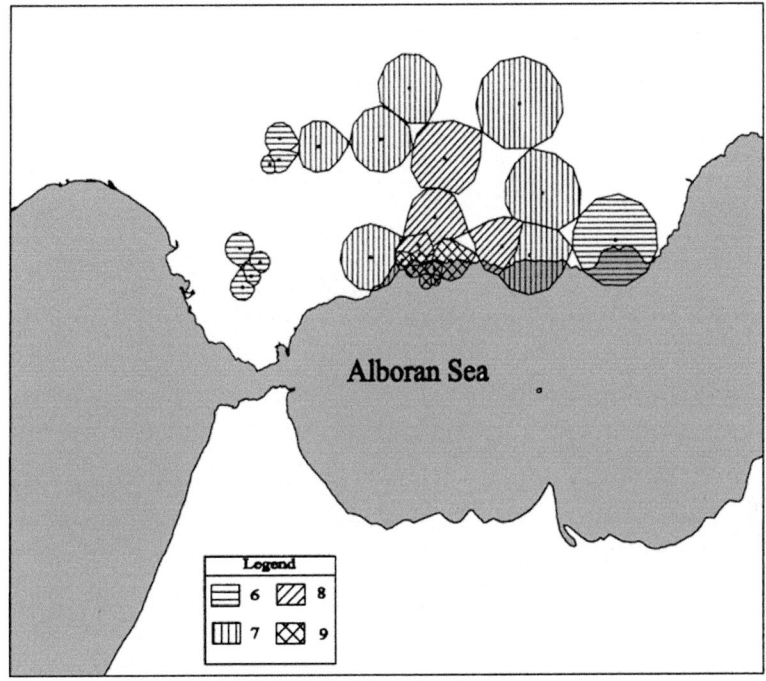

Figure 5
Macroseismic plane of the 10-09-1680 earthquake (MEZCUA, 1982).

construction of intensity virtual areas (TERMINI *et al.*, 2005c) that complete the intensity distribution of the earthquake (Fig. 6).

This last macroseismic plane represents the intensity distribution of reference for the analytical determination of epicenter and principal attenuation directions, whose reliability can be deduced, respectively, through their consistency with the maximum shaking area and the geometry of structural lineaments of the region (BOTTARI *et al.*, 1984; TERAMO *et al.*, 1995a, b, 1996). In Figure 7 and in Table 1 and in Figure 8 and Table 2, the differences between the observed epicenters and the calculated ones highlight with reference, respectively, to historical and recorded earthquakes, a good reliability of results and, in some cases, also a specific refining of previous knowledge.

Moreover, the determination of the corresponding attenuation directions, reported in Figure 9 with reference to the whole set of analysed seismic events, allows different sets to be defined that can be considered homogeneous in terms of attenuation directions and therefore a suitable instrument to try in characterizing a seismic zonation, considering the lack of a territory subdivision into seismogenic zones.

Other applications of this procedure have been adopted with reference to the other available intensity maps of earthquakes of southern Spain, with specific

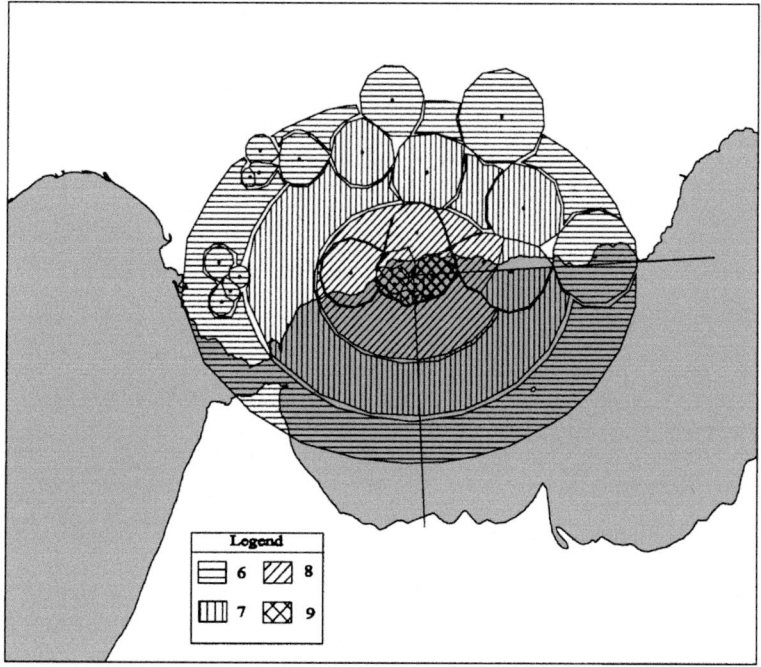

Figure 6
Intensity virtual areas of the 10-09-1680 earthquake.

evaluations of the correct intensity level to be attributed to the points of the corresponding intensity maps.

4. Discussion and Conclusions

The modelling of the distribution of macroseismic intensity of historical and recorded earthquakes in southern Spain, accomplished through procedures already applied in southern Italy and Sicily with appreciable results, allows in-depth studies pertaining to seismicity of the region in study.

The elements that have characterized all the analysis and for which suitable configurations have been depicted, are essentially referable to the lack of reliable documents relative to the historical earthquakes, some of which occurred along the coast; the consequent reduced number of intensity observed points of the corresponding intensity maps and their irregular distribution; the objective difficulty in individualizing attenuation directions consistent with the actual geometry of structural lineaments. The adopted procedure, in spite of all these uncertainties, allows specific modelling of intensity distribution whose reliability derives from the

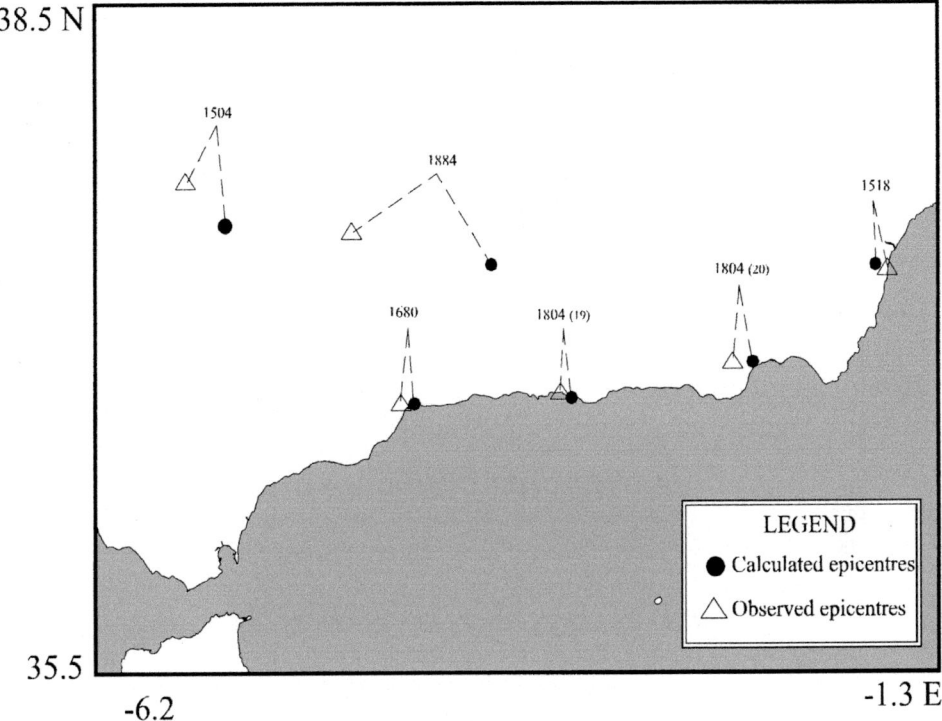

Figure 7
Comparison between the observed and the calculated epicenters of historical earthquakes.

consistency of the calculated epicenters and attenuation directions respectively, with the areas of maximum shaking and the structural lineaments. This consistency, with reference to earthquakes which occurred along the coast, has been reached through the use of intensity virtual areas in zones close to the sea, in which the lack of macroseismic information makes the intensity modelling difficult for the determination of the macroseismic parameters.

In this context, it is necessary to highlight that the complex system of structural lineaments of the region does not allow a certain and efficient individualization of several homogeneous sets of attenuation directions, that can be identified as a first attempt to characterize some seismogenic zones for seismic hazard evaluation. In particular, for some areas such as the Granada basin or the Malaga area, the calculated attenuation directions, mainly oriented respectively towards N-NW/S-SE and W-E (Fig. 9), are consistent with the structural lineaments of the same area (Fig. 2).

Actually, it would be desirable if more specific evaluations regarding the structural framework for the entire region were carried out with the aim of correlating each earthquake to the corresponding fault, taking into account the

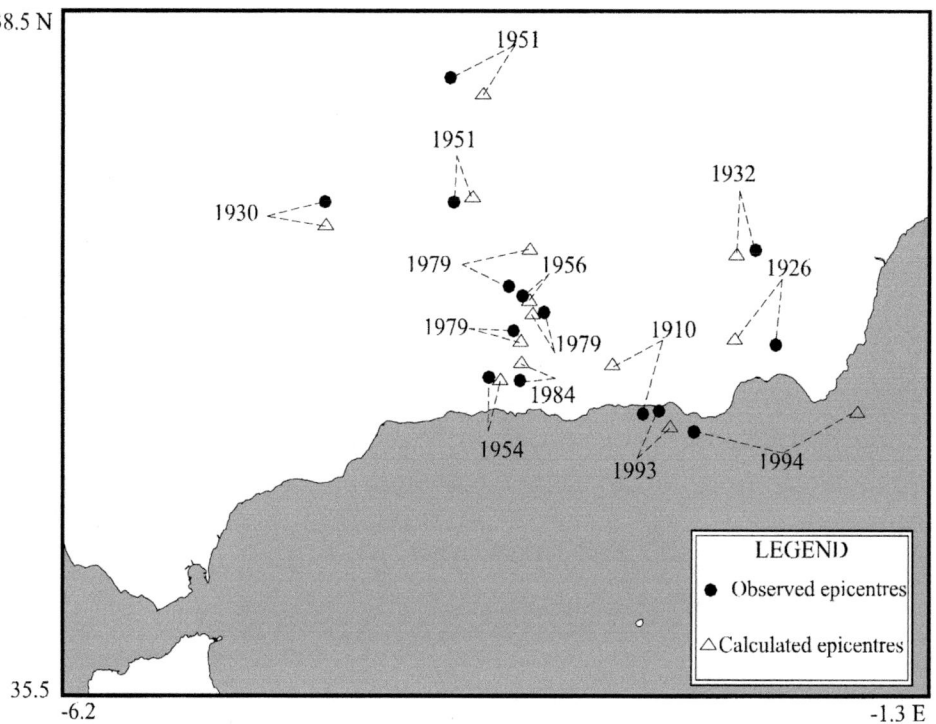

Figure 8
Comparison between the observed and the calculated epicenters of recorded earthquakes.

intensity distribution modelling of the earthquakes which occurred at very close sites, to configure a reliable approach to the individualization of a number of seismogenic zones.

Acknowledgements

The authors warmly thank Prof. Mario Boccaletti for his contribution on the geostructural framework; Dr. Manuel Espinar, of the Instituto Andaluz de Geofisica, Universidad de Granada, for his useful historical descriptions.

Dr. T. Tuvè collaborated with the composition of this paper within studies treating the historical seismicity of Andalusia which was the theme of her Doctorate thesis on Geophysics for Environment and Territory.

REFERENCES

BLANCO, M.J. and SPAKMAN, W. (1993), *The P-wave Velocity Structure in the Mantle below the Iberian Peninsula: Evidence for Subducted Lithosphere below Southern Spain,* Tectonophysics *221*, 13–34.

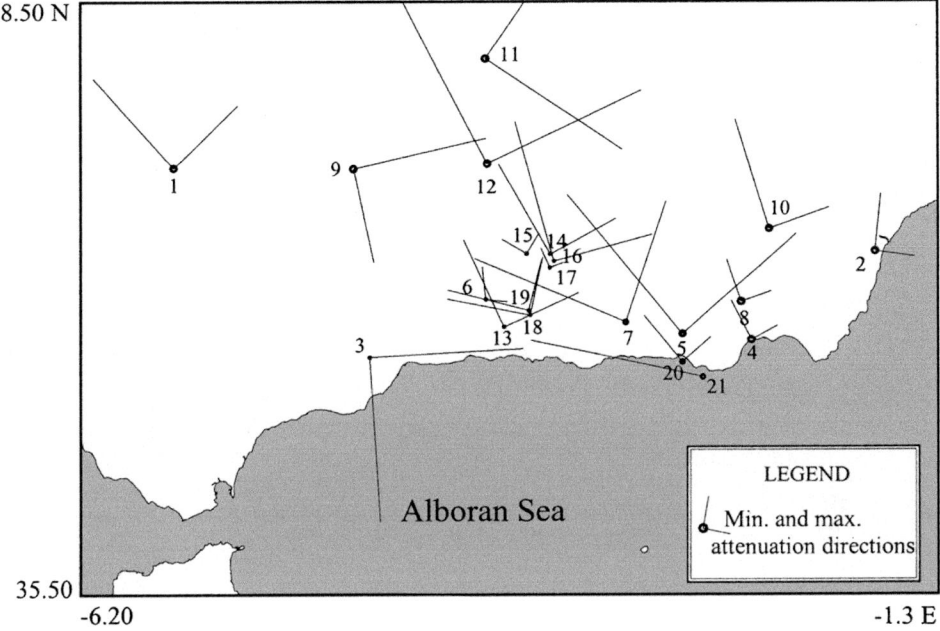

Figure 9
Maximum and minimum attenuation directions of the analyzed earthquake set.

BOCCALETTI, M., PAPANI, G., GELATI, R., RODRIGUEZ-FERNANDEZ, J., GARRIDO LOPEZ, A.C., and SANZ DE GALDEANO, C. (1987), *Neogene-Quaternary Sedimentary-tectonic Evolution of the Betic Cordillera*, Acta Naturalia de l'*Ateneo Parmense 23*, 179–200.

BOTTARI, A., FEDERICO, B., and LO GIUDICE, E. (1984), *The Correlation between the Macroseismic Attenuation Trend and the Geostructural Framework: The Calabro-Peloritanian Arc as an Example*, Tectonophysics *108*, 33–49.

GALINDO-ZALDÍVAR, J., JABALOY, A., SERRANO, I., MORALES, J., GONZÁLEZ-LODEIRO, F. and TORCAL, F. (1999), *Recent and Present-day Stresses in the Granada Basin (Betic Cordilleras): Example of a Late Miocene-present-day Extensional Basin in a Convergent Plate Boundary*, Tectonics *18*, 686–702.

GONZÁLEZ-LODEIRO, F., ALDAYA, F., GALINDO-ZALDIVAR, J., and JABALOY, A. (1996), *Superposition of Extensional Detachments during the Neogene in the Internal Zones of the Betic Cordilleras*, Geol. Rundschare *85*, 350–362.

IBAÑEZ, J.M, BOTTARI, C, ESQUIVEL, J.A, MORALES, J., and ALGUACIL, G. (1999), *Magnitude-intensity Relations for Instrumental and Historical (IX-XIX Century) Earthquakes in South Spain. Tectonical and Seismic Hazard Implications*, Bull. Seismol. Soc. Am., still in preparation.

LONERGAN, L. and WHITE, N., (1997), *Origin of the Betic-Rift Mountain Belt*, Tectonic *16*, 504–522.

LOPEZ-CASADO, C., MOLINA-PALACIOS, S., DELGRADO J., and PELAEZ, J.A. (1999), *Attenuation of Intensity with Epicentral Distance in the Iberian Penisula*, still in preparation

MEZCUA, J. (1982), *Catalogo general de isosistas de la Penisula Iberica*, Publicaciòn 202 IGN, Madrid.

MORALES, J., SERRANO, I., JABALOY, A., GALINDO-ZALDIVAR, J., ZHAO, D., TORCAL, F., VIDAL, F., and GONZÁLEZ-LODEIRO, F. (1999), *Active Continental Subduction beneath the Betic Cordillera and the Alborán Sea*, Geology *27*, 735–738.

SANCHEZ NAVARRO NEUMAN, M. (1921), *Lista de terremotos mas notables sentidos en la Penisula Iberica (a.C. 1917)*, Est. Sism. De Cartuja, 11–65.

SANZ DE GALDEANO, C., LOPEZ CASADO, C., DELGADO, J., and PEINADO, M.A. (1995), *Shallow Seismicity and Active Faults in the Betic Cordillera. A Preliminary Approach to Seismic Sources Associated with Specific Faults*, Tectonophysics *248*, 293–302.

SEBER, D., BARAZANGI, M., IBERBRAHIM, A., and DEMNATI (1996), *Geophysical Evidence for Lithospheric Delamination beneath the Alborán Sea and Rif-Betic Mountains*, Nature *379*, 785–790.

TERAMO, A., STILLITANI, E., and BOTTARI, A. (1995a), *Anisotropic Characterization of Macroseismic Fields*, Natural Hazard *11*, 223–245.

TERAMO, A., STILLITANI, E., and BOTTARI, A. (1995b), *An Anisotropic Attenuation Law of the Macroseismic Intensity*, Natural Hazard *11*, 203–221.

TERAMO, A., TERMINI, D., STILLITANI, E., and BOTTARI, A. (1996), *The Determination of Epicentre by a Vectorial Modelling of Macroseismic Intensity Distribution*, Natural Hazard *13*, 101–117.

TERMINI, D., BOTTARI, A., TERAMO, A., and TUVÈ, T. (2005a), *On the Observed Intensity Filtering in the Anisotropic Distribution Modelling of Macroseismic Intensity*. Pure and Appl. Geophys. *162*, 683–697

TERMINI, D., TERAMO, A., BOTTARI, C., and TUVÈ, T. (2005b) *An Anisotropic Attenuation Law of Macroseismic Intensity Performed on Virtual Intensity Distribution of Seismogenic Zones*. Pure Appl. Geophys. *162*, 707–714.

TERMINI, D., BOTTARI, A., TERAMO, A., and BOTTARI, C. (2005c), *On the Intensity Virtual Area Characterization in the Intensity Distribution Modelling by Macroseismic Planes*, Pure Appl. Geophys. *162*, 699–705.

VISSERS, R.L.M., PLATT, J.P., and VAN DER WALD. (1995), *Late Orogenic Extension of the Betic Cordillera and the Alboran domain: A Lithospheric View*, Tectonics *14*, 786–803.

WEIJERMARS, R. (1985), *Uplift and Subsidence History of the Alboran Basin and a Profile of the Alboran Diapir (W-Mediterranean)*, Geol. Mijnbouw *64*, 349–356.

(Received September 5, 2001, revised December 6, 2001, accepted January 10, 2002)

 To access this journal online:
http://www.birkhauser.ch

Pure appl. geophys. 162 (2005) 761–765
0033–4553/05/040761–5
DOI 10.1007/s00024-004-2639-6

© Birkhäuser Verlag, Basel, 2005

▌**Pure and Applied Geophysics**

Ancient Constructions as Markers of Tectonic Deformation and Strong Seismic Motions

Carla Bottari[1]

Abstract — Many ancient structures such as temples were constructed on the basis of a very strict plan and excellent workmanship. For this reason, even their slight deformation due to various effects (ground instability, earthquake oscillations, etc.) can be identified, and it is possible to discriminate between different types of deformation due to earthquakes and those due to other natural causes or to anthropogenic effects. Two study cases are presented here: the Propylaia on the Acropolis and the temple of Hephaistos (Hephaisteion) in the Agora at Athens. In both buildings deformation was produced by earthquakes.

Key words: Earthquake, deformation.

1. Introduction

Various ancient buildings have suffered damage over the centuries, chiefly due to human intervention such as fire, military and wartime destruction, explosions etc., but also due to natural causes such as falling rocks, thunderstorms, earthquakes etc. Unfortunately, only a few of these structures are still standing and preserve traces of seismic deformation, while other buildings that were completely destroyed in the rest or have been recently restored do not preserve any evidence of seismic deformation.

In Athens, the historical and instrumental seismic data indicated that it is an area of low-seismicity. In fact only a few earthquakes are documented in the period between the fifth century B.C. and the seventeenth century A.D. The earliest known earthquake occurred in 427 B.C., and from historical reports it seems to have caused minor damage to the Acropolis in Athens. Some time later, before the Frankish occupation (A.D. 1200), a stronger earthquake struck the city producing dislocation in the NE corner of the Parthenon (KORRES, 1996). Many other earthquakes have struck this region over the course of years but only a few were recorded, while others, such as the earthquake that caused damage in the temple of Hephaistos, have recently been discovered (GALANOPOULOS, 1956).

---

[1] Osservatoriosismologico, Università ʻdi Messina, Via Osservatorio, 4 – 98121 Messina, Italy
E-mail: cbottari@unime.it

In this article we summarize some of the earthquake deformations observed in ancient monuments in Athens estabilishing that the town was not aseismic in the past and that through the study of earthquake damage we can improve our knowledge of past seismic activity.

2. Cases in Study

2.1. Hephaisteion

This Doric temple was built by the Greeks between 449 and 415 B.C. on the *Kolonos Agoraios*, the 66m high Hill of the Agora. It is known as Theseion although the temple was probably dedicated to Hephaistos, the God of Metalworking, and it is the best preserved temple in the Greek world. It is a rectangular building measuring 31.8 m long and 13.7 m wide, with 13 columns on the long sides and 6 on the short sides. The superstructure is made of Pentelic and Parian marble while the foundations are built from limestone. The lowest step of the temple, built from

Figure 1
Sinusoidal offset of the column drums indicative of seismic deformation in the Hephaisteion (Theseion), Athens according to GALANOPOULOS (1956).

poros limestone, stands directly on dressed bedrock with the exception of the southwest and southeast corners, which stand on foundation courses of a different kind of poros (DINSMOOR, 1941). During Byzantine times (around the seventh century A.D.), the temple of Hephaistos was transformed into a church and the entrance was moved to the west side. Later it was used as a cemetery.

According to GALANOPOULOS (1956) the temple suffered an earthquake, which probably occurred after the Byzantine restorations (Stiros' personal communication). The seismic event was not strong enough to cause the collapse of the building, however the southern side of the temple suffered a deformation that is still preserved today (Fig. 1). The sinusoidal offset of the column drums is indicative of seismic deformation. This was brought on by the different mechanical behavior of the two types of limestone in the substructure during the seismic shaking.

2.2. The Propylaia

The gate building of the Acropolis was designed by the architect Mnesikles and built shortly after the construction of the Parthenon, between 437–432 B.C. Two different types of stone were used: one Eleusinian blue and the other Pentelic white. It was intended as a façade for the Acropolis and the end-point of the most important festival in ancient Athens, the Panathenaia. The building was never finished and it had three different kinds of columns, small Doric, large Doric and Ionic columns. Six large Doric columns on the front were flanked by two asymmetrical porches. A wide

Figure 2
Characteristic rotation in the column drums of the Propylaia after the 1999 Athens earthquake.

passageway with a central ramp flanked by six Ionic columns, three on each side, was positioned between the large Doric columns (Fig. 2; LAWRENCE, 1996). From the twelfth to the fifteenth century A.D. the building was transformed to host the Dukes of Athens and at the same time a quadrangular Frankish tower was added at the east wing to reinforce the defence of the Acropolis. Subsequently the military superstructure was demolished in the nineteenth century.

On September 7th, 1999 an earthquake caused an offset in the drums of the columns on the central ramp. This seismic event was not stronger than the average earthquake recorded in Athens — in fact the calculated M was around 5.5–5.9 — but the peak ground acceleration was high enough (0.30 g) to produce a rotation of the drums (Fig. 2).

3. Conclusion

The offset of column drums in ancient Greek or Roman buildings, when their base is stable, is indicative of seismic oscillations produced by high acceleration seismic motion. The ground acceleration of the 1999 earthquake (with its epicenter about 15 km from the Acropolis) was high enough to produce a rotation of the drums of the Propylaia while the previous studies on seismic risk for Athens had estimated a maximum acceleration of 0.16 g (ANASTASIADIS et al., 1990). Another very strong earthquake producing damage in Athens was the 1981 Corinth earthquake, with $M_s = 6.7$, PGA $= 0.29$ g and with an epicentral distance of about 77 km (ANASTASIADIS et al., 1990). This earthquake produced no damage in the nearby archaeological site of Corinth while in the Parthenon at Athens it opened joints (around 1 cm between the blocks) in the western part of the S wall (KORRES, 1996).

In conclusion, through the study of deformation of the ancient buildings we can improve our knowledge of the seismic history of an area under study and in the specific case of Athens we can affirm it had experienced earthquakes throughout its history, as the above-mentioned evidence shows. Furthermore, the study of seismic deformation provides an opportunity to collect additional qualitative and quantitative information that can serve as a marker in the studies on the seismic behavior of monuments.

Acknowledgements

I thank Prof. Stathis Stiros for the highly-valued information and for the opportunity to write this paper. I also thank Justin St. P. Walsh for reviewing the English text and archaeological data.

REFERENCES

ANASTASIADIS, An., M. DEMOSTHENOUS, Ch. KARAKOSTAS, N. KLIMIS, B. LEKIDIS, B. MARGARIS, Ch. PAPAIOANNOU, PAPAZACHOS, C. and THEODULIDIS N. (1990), *The Athens (Greece) Earthquake of September 7, 1999: Preliminary Report on Strong Motion Data and Structural Response.* http://www.itsak.gr/report.html

DINSMOOR, W. B. (1941), *Observations on the Hephaisteion,* Hesperia Suppl. 5.

GALANOPOULOS, A. (1956), *The Seismic Risk at Athens,* Praktika Akadimias Athinon *31,* 464–472 (in Greek).

KORRES, M. (1996), *Seismic damage to the monuments of the Athenian Acropolis.* In: *Archaeoseismology* (Stiros, S. and Jones, R., eds.) Fitch Laboratory Occasional Paper 7. British School at Athens: 69–74.

LAWRENCE, A. W. (1996), *Greek Architecture,* Pelican History of Art, Yale UP: New Haven, Conn.

(Received July 25, 2003, accepted February 12, 2004)

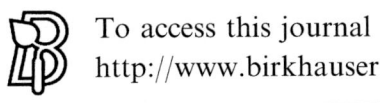

To access this journal online:
http://www.birkhauser.ch

Pure appl. geophys. 162 (2005) 767–782
0033–4553/05/040767–15
DOI 10.1007/s00024-004-2640-0

❚ Pure and Applied Geophysics

Methodological Considerations for the Evaluation of Seismic Risk on Road Network

ANTONINO D'ANDREA[1], SALVATORE CAFISO[2], and ANTONIO CONDORELLI[2]

Abstract—The recent earthquakes in California and Japan have shown the fundamental role that the road infrastructure plays in emergencies. In fact, only the maintenance of a sufficient level of efficiency can help to quickly reach the affected areas and thus avoid further serious consequences. The necessity of guaranteeing the functionality of the transport network during seismic events therefore requires seismic risk planning extended also to the road infrastructures in order to support the management of post-earthquake emergencies. Analogously it is fundamentally important to have analysis instruments of the road system able to preventatively evaluate the effects of earthquakes in order to identify possible emergencies, therefore preparing a program of intervention to reduce seismic risk on road networks. This paper proposes a methodology for the evaluation of seismic risk of road infrastructures according to the following points:

• Study of seismic hazard of the site for the definition of a seismic scenario using attenuation models in relation to historical seismology and the geological and tectonic characteristics of the territory;
• Analysis of the direct exposure connected to the probability of the presence of road users on the different parts of the network directly exposed to the seismic event;
• Analysis of the indirect exposure relative to the distribution of the population and the infrastructures for which post-earthquake accessibility must be guaranteed;
• Evaluation of the functional vulnerability in relation to the potential replaceability of damaged stretches considering network configuration and geometrical characteristics;
• Evaluation of structural vulnerability of the stretch correlated to the characteristics (structural, mechanical, technological, etc.) of the different components (bridges, embankments, trenches, tunnels) that make up the stretches obtained by the use of correctly elaborated tables for each component.

The determination of global risk indexes of the single stretches and of the network, evaluated by means of a relationship between the ascertained parameters derived from the investigation of the previous points, provides the necessary information for the definition of mitigation measures to reduce the risk and for management planning before and after disaster. The proposed methodology, which has already been applied to a restricted area, is currently being applied to the province of Catania (Sicily, Italy), which is one of the geographical regions of highest seismic risk in Europe, and its future extension to all of eastern Sicily is foreseeable.

Key words: Road, earthquake, seismic risk, natural hazard, vulnerability.

[1] Dipartimento di Costruzioni e Tecnologie Avanzate, Facoltà di Ingegneria, Università di Messina, Contrada Sperone, Vill.S.Agata, 98166, Messina, Italy.
E-mail: dandrea@ingegneria.unime.it
[2] Dipartimento di Ingegneria Civile ed Ambientale, Facoltà di Ingegneria, Università di Catania, V.le A.Doria 6, 95125 Catania, Italy.

1. Introduction

Seismic events, as can be seen from the history of Italy, are natural disasters that often have catastrophic consequences for the population and all the anthropic systems. With the aim of setting up prevention measures and anti-seismic retrofitting, methodologies have been introduced for the evaluation of seismic risk for buildings in order to define expected damage following a quake on a given area and to identify all the elements at high risk. In this risk evaluation an important component often has been overlooked, or at least, not considered as being of great importance: the transportation system.

The recent Californian (1989 and 1994) and Japanese (Kobe, 1995) earthquakes have tragically shown the essential role that transportation networks have in an emergency condition after a seismic event.

Only if the transportation network remains efficient, will assistance quickly and easily reach the stricken areas, avoiding or limiting subsequent consequences that could cause damage of a comparable extent to those directly induced by the earthquake.

In the case of Kobe, for example, it was ascertained that the damage from fires indirectly caused by the quake determined damage comparable to that caused directly by the quake, in as much as the interruption of the access ways prevented emergency services from reaching the devastated areas for many hours. Therefore, other than the *direct* exposure of a seismic event, the interruption of the transportation network causes an *indirect* exposure of the population living in stricken areas.

In a global evaluation of seismic risk it is fundamental to consider the *direct* exposure of the users of the transportation network beyond the "classical" one of the resident population of the urban areas. On some road infrastructures, in fact during much of the day, a great number of people are exposed to risk as well as those who are inside buildings. It should be remembered that in the Loma Prieta earthquake (1989) the collapse of the busy Cypress Street viaducts brought about many of the victims of the earthquake. For all these reasons, of which, unfortunately, we have understood the importance only after tragic events such as this, the network of road infrastructures is today considered one of the most important *lifeline* systems.

In this context this study proposes a methodology for the evaluation of the seismic risk on rural road system infrastructures. This procedure is based on three different phases that converge in the final definition of the risk:

- Study of seismic hazard;
- Analysis of seismic exposure;
- Evaluation of seismic vulnerability.

The methodology has been planned through opportune representative indicators for its Geographic Information System (GIS) application.

2. Study of Seismic Hazard

The first phase of establishing a procedure for seismic risk evaluation of a given area is studying seismic hazard. This term defines the *probability of occurrence of a seismic event of each intensity, in a definite area and in a definite period of time.*

The evaluation of seismic hazard of an area is based on the study of historic seismology and on the analysis of the geological seismologic and seismogenetic characteristics of the site.

The historical studies are aimed at the definition of the principal geophysical characteristics (epicenter, magnitude, ground acceleration, etc.) of seismic events that have struck the area under examination. As the scientific community believes that an area hit by a seismic event can be impacted again by a similar event, these investigations are invaluable to characterize an adequate earthquake scenario for the studied area.

By the analysis of seismological and seismogenetical characteristics of the area it is possible to identify, on a vast scale, active or potentially active faults. By the study of geological and geomechanical characteristics of the land it is possible to evaluate, for each site, the various reaction mechanisms to a seismic shock of the different geological structures that are present, in such a way as to predict the effects of a certain earthquake in terms of horizontal force, acceleration, etc.

Once the definition of the geophysical characteristics of an earthquake scenario have been completed, an adequate model of attenuation therefore needs to be chosen according to the geophysical and geomechanical characteristics of the site in order to apply it to the studied areas to build up a map of the seismic hazard on the basis of a representative indicator.

Considering the GIS application of the methodology that provides a ready possibility to create several different simulations (D'ANDREA *et al.*, 1996, 1997), it was decided to define the seismic hazard by way of the "earthquake scenario." Therefore, as a representative indicator of the seismic hazard, the Peak Ground Acceleration (PGA) was used so as to allow the management at the territorial level of very significant synthetic parameters. The spatial distribution of PGA can be evaluated through opportune laws of attenuation which correlate the distance of the epicenter, the magnitude of the earthquake, the local constitution of the land and other parameters.

In the first experimental application in the province of Catania (Italy), the spatial PGA distribution expected was evaluated by an attenuation law (SABETTA and PUGLIESE, 1987, 1996) opportunely set for the Italian territory. An earthquake similar to that which occurred in the same zone in 1693 was used, hypothesizing an epicenter at sea in the Ibleo-Maltese fault at about 9 km from the coast with a magnitude of 7.3 on the Richter scale. The law that was used is very sensitive not only for the distance of the epicenter but also for the different properties of

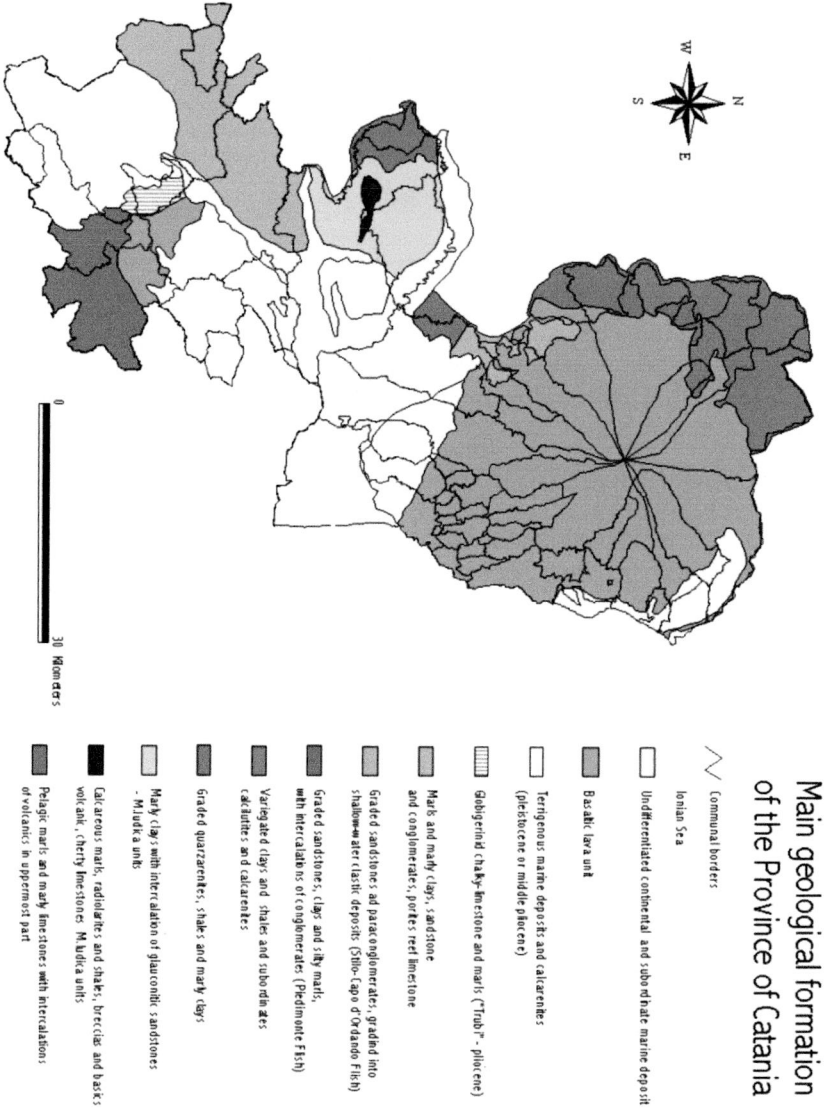

Figure 1
Main geological formations of the Province of Catania.

transmission of the seismic waves of the various geological formations (Figs. 1, 2).
This simplified assumption for the seismic hazard produces acceptable results with
respect to the maximum expected earthquake, and could be repeated for a different
one to obtain a new scenario.

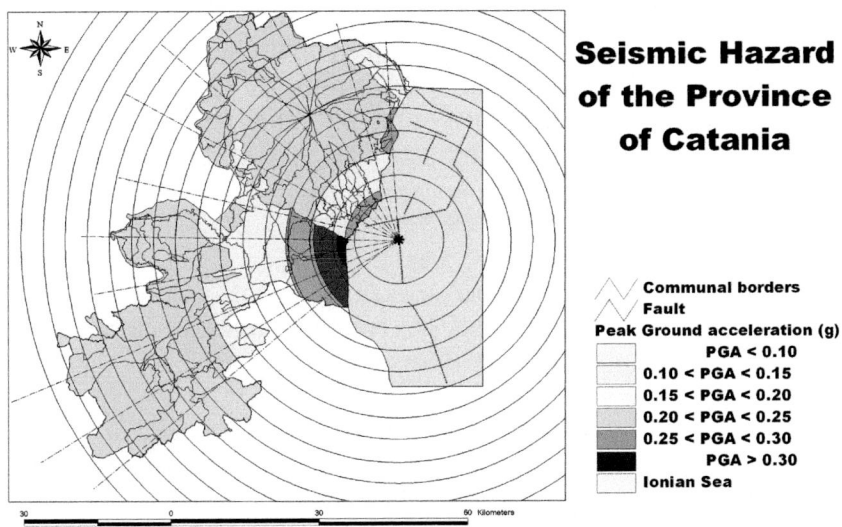

Figure 2
Seismic hazard of the Province of Catania.

3. Analysis of Seismic Exposure

The term *seismic exposure* defines *the extension, the quantity, the quality of the different anthropic elements that make up the territorial context (population, buildings, system of infrastructures...) whose conditions and/or whose operation could be altered, modified in any way or damaged in a seismic event.*

The analysis of seismic exposure represents the second fundamental step after the study of seismic hazard. This, in fact, identifies all the anthropic elements that, being in an area of ascertained seismic hazard, could be susceptible to damage directly or indirectly caused by the earthquake (CAFISO *et al.*, 1999).

Here we distinguish two kinds of seismic exposure:

- **direct** exposure, concerning the consequences that could be linked directly to the users of the transportation network during an earthquake that damages the infrastructures;
- **indirect** exposure, in which indirect damage caused by the partial inability of the transportation network is considered.

As regards *direct exposure*, it is clear that it will not be possible to refer to the population as a static entity. On the contrary it will be necessary to consider the population as a dynamic element that assumes, continually, in the area, concentrations and a different distribution using the transportation system for their trips. In this study Average Annual Daily Traffic (AADT) has been used as a representative indicator of direct seismic exposure on each stretch of the road network. The knowledge of AADT allows the construction of a map of the direct seismic exposure of the population of users in each stretch of road.

The analysis of *indirect seismic exposure* consists in a study of estimations of damage or injury to the population that can result from a road network or of a part of it that does not function correctly. It is, obviously, a study of exposure because our principal object is the analysis of damage or injury for people as an indirect consequence of breakdowns in or the bad efficiency of road networks. In the emergency phases that follow a seismic event the transportation network has the task of making assistance accessible to the stricken area so that aid can be quick and efficient. If road infrastructure efficiency has been compromised due to the effects of the earthquake, to reach the stricken areas would be impossible or quite slow and difficult.

It is evident that the damage caused by this kind of condition is different according to the dimensions and characteristics of the city that has been hit by the earthquake. Each town, in fact, has to be considered as a generator of demand for assistance that is proportional to the resident population. Indirect exposure of each stretch of the network increases with the growth of the resident population in the towns linked to it.

Another fundamental aspect is identification, position of the strategic structures existing in the towns to be used for civil protection: hospitals, military barracks, fire and police stations, and rallying points, etc. Therefore we have to consider and evaluate each town as a potential provider of assistance for the surrounding territory. However, the greater the potential assistance of the town, the greater would be the indirect seismic exposure of the roads of the network linked to them due to the greater potential that they can provide.

The correct way to analyze the indirect seismic exposure of road system infrastructures is to consider each town of the territory as a potential generator of demand or provider of assistance. If the potential for assistance is great, then also the indirect exposure of roads linked to the town would be great.

It is necessary, after the definition of indirect exposure, to evaluate the level of exposure of each stretch of the network using a specific Origin/Destination (O/D) study dedicated to the mobility of the emergency services. The points must be referred to the principal towns, or to the territorial areas in which strategic structures for civil protection are placed and to the towns that have to depend on the principal towns for assistance. The level of exposure, therefore, will be proportional to the population involved in collapses (Pc) and that will have to be helped using the studied stretches. The values of Pc in Italy have been defined by the Civil Defence Department (PCM-DPC) using the studies of seismic risk in urban areas. The analysis of indirect exposure conducted on the roads that have the primary function of linking nodes (towns, cities and urban centers) will allow the identification of lifelines on which further studies of risk will be conduced.

The map of Figure 3 proposes a preliminary classification of urban centers in order to evaluate indirect seismic exposure.

4. Evaluation of Seismic Vulnerability

Seismic vulnerability is defined as *the propensity of an element, simple or complex, to suffer damage, collapse or modification during a seismic event.* As we know, seismic vulnerability is an intrinsic characteristic of each construction, that is independent from any kind of external factor. For example, the vulnerability of a bridge depends on the construction technologies adopted, on the materials employed, on its structural configuration, on its age, on its state of conservation, on the quality of the original project etc. All these factors are independent from the localization of the object and from the probability that a seismic event can take place there, which has been already evaluated in the study of seismic hazard.

Extending the concept referred to a single construction to a total system, the vulnerability can be intended as a measure of propensity to lose or reduce its efficiency in carrying out its function.

Road network infrastructures are composed of different stretches that converge in or diverge from nodes: each stretch is assumed to be composed of a series of components with homogeneous characteristics (bridges, embankments, trenches, tunnels, etc.)

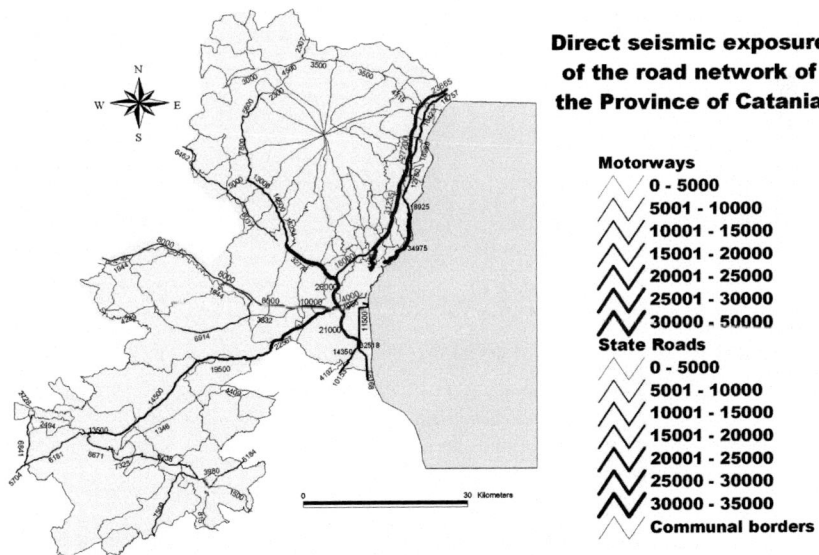

Figure 3
Direct seismic exposure of the road network of the Province of Catania [AADT].

For a system that is so complex and full of interrelations among the elements that constitute it, it is not possible to have a definition of vulnerability with a value that is exclusively structural, however it is necessary to extend the concept to encompass a functional value. For this aim it is essential to identify the following levels of study:

- Evaluation of Structural Factors (Re, G) of the components;
- Evaluations of Functional Vulnerability (Vf) of the different stretches of the network.

The first phase comprises the identification of the components that constitute the stretches of the network and in the analysis of their structural and constructive characteristics. The components have been characterized in:

- Bridges and viaducts;
- Embankments;
- Trenches;
- Galleries and tunnels.

Table 1a

Factors for the evaluation of seismic vulnerability index of bridges and viaducts

Element	Min. Vulnerability		Max. Vulnerability
Design criteria	Constructed according to antiseismic criteria		Constructed without antiseismic criteria
Construction type	Continuous structures		Discontinuous structures
Regularity of geometry and rigidity	Regular structures		Irregular structures
Pier type	Single		Multiple
Abutment height	Low		High
Soil-foundation system	good		Bad and with liquefaction problems
Condition of the construction	Good state of conservation		Bad state of conservation
Alignment	Low angles of deviation		High angles of deviation
Type of bearing support	If longitudinal and transversal movement is allowed and there are systems to prevent girder fall		Simple support
Expansion joints	None	Present with antiseismic criteria	Present with a short base
Building material	Steel	Reinforced concrete	Masonry

Table 1b

Factors for the evaluation index of seismic vulnerability of embankments

Element	Min. Vulnerability			Max. Vulnerability
Design criteria	Constructed according to antiseismic criteria			Constructed without antiseismic criteria
Height	Low			High
Geometrical condition of the site	Low inclination of ground	High inclination of ground with antiseismic wall	High inclination of ground with no antiseismic wall	High inclination of ground without wall
Embankment geometry	Slope inclination < 2/3			Slope inclination > 2/3
Soil support characteristics	Good			Bad
Condition of the structure	Good state of conservation			Bad state of conservation
Slope protection	Presence of slope protection			No slope protection

Table 1c

Factors for the evaluation index of seismic vulnerability of trenches

Element	Min. Vulnerability		Max. Vulnerability
Design criteria	Constructed according to antiseismic criteria		Constructed without antiseismic criteria
Geological and geometric condition of slope	Dynamic safety coefficient Fs ≥ 1.3		Dynamic safety coefficient Fs < 1.3
Length – height	Trenches low and short		Trenches long and high
Rock fall	Impossible	Possibility of rock fall with slope protection	Possibility of rock fall without slope protection
Retaining structures	built according to antiseismic laws		Not built according to antiseismic laws

For each of these typologies it is possible to construct an evaluation table of structural vulnerability on the basis of the most important elements involved in the phenomenon, for example: design criteria (without specific data we must refer to the antiseismic laws in force during the period of construction), state of conservation, geometrical dimensions, structural characteristics and those of soil, etc. In the following tables (Tables 1a-1d) we supply a selection of the principal elements that require consideration to evaluate the indicator of Vs (structural vulnerability) for

Table 1d

Factors for the evaluation index of seismic vulnerability of tunnels

Element	Min. Vulnerability		Max. Vulnerability
Design criteria	Constructed according to antiseismic criteria		Constructed without antiseismic criteria
Geostructural condition of the mass	Good: no earth pressure – absence of discontinuity		Bad: earth pressure – presence of landslides and discontinuities
Deformation joint	Presence of deformation joints		Absence of deformation joints
Location	Deep tunnel	Superficial tunnel	Galleries
Section area	Small		Large
Section type	Closed section		Open section
Condition of the structure	Good state of conservation		Bad state of conservation

each component. In particular, in the case of bridges and viaducts literature provides definite criteria for a quantitative evaluation (COOPER and VANDEPOL, 1991; BUCKLE and KIM, 1995; ZÜLFIKAR and YÜZÜGÜLLÜ, 1995). It is therefore possible to obtain a breakdown of each stretch in homogeneous components characterized by an indicator of vulnerability (AIPCR, 1996; ARENA *et al.*, 1997; D'ANDREA, 1985; D'ANDREA and BOSURGI, 1999; AMBRASEYS and SRBULOV, 1995).

Table 2a

Evaluation table of parameter G (Geometry)

Classification	G
Local rural roads	1
Secondary rural roads	
Principle rural roads	
Highways	< 1

Table 2b

Evaluation table of Re (replaceability)

Number of routes available	Re
Only one	1
More than one, but near	
More than one located in distant areas	< 1

To characterize a stretch with only one indicator of vulnerability, many criteria of aggregation could be proposed, among which one of the simplest and of immediate application consists in giving to the entire stretch the maximum value from the indicators of structural vulnerability of the components that make it up.

The vulnerability of a road network is not just the seismic resistance of each stretch, but also and above all the global working of the system even after a earthquake. It is now obvious that not all the stretches have the same role and importance for strategic movement in the territory. The evaluation of functional factors that can reduce Vs therefore depends on the possibility of alternatives offered by the network replaceability and by the distance between alternative links (functional factor Re). The level of replaceability is linked to the possibility offered by the network of guaranteeing alternative routes in case of damage to some of them. The proximity of two stretches of alternative routes makes the probability greater that an earthquake damages them both. On the other hand, the best geometrical characteristics of the links of the network (width section, alignment, etc.) generally make conditions more favorable for transit even in the case of partial damage (functional factor G). Functional vulnerability can be evaluated with the criteria shown in Tables 2a, 2b. The functional aspects of the stretches of network expressed as G*Re can be considered as an adequate reduction factor of structural vulnerability.

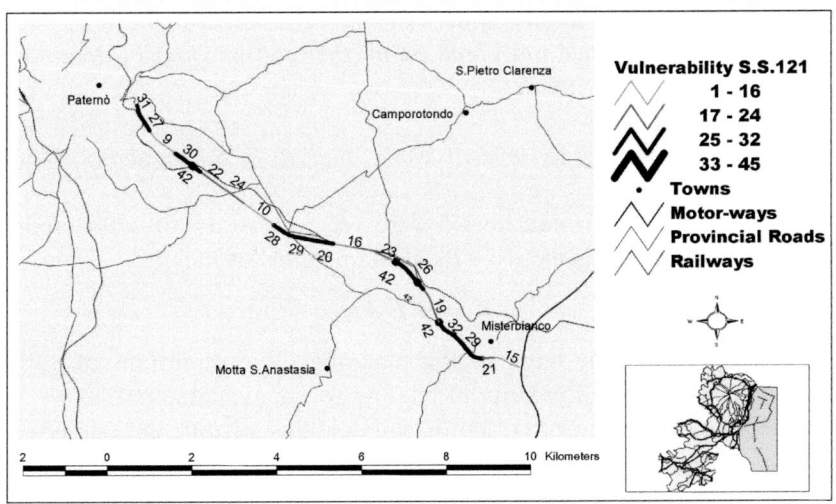

Figure 4
Structural vulnerability (Vs) of the components of State Road 121.

The total vulnerability (Vt) of each stretch derives from the aggregation of functional factors (Re, G), given from the analysis of the above-mentioned network (Tables 2a–2b), and structural vulnerability (Vs), from the components of the stretch that, preliminarily, can be deduced with the procedure developed based on the criteria of Tables 1a–1d. In this way it will be possible to represent stretches of equal structural vulnerability with different total vulnerability on the basis of the presence of more favorable geometrical conditions or of the existence or not of alternatives.

$$Vt = Vs^*(G^*\text{Re}). \tag{a}$$

At the present the proposed methodology was applied for the evaluation of functional vulnerability and for structural vulnerability exclusively for the Misterbianco–Paternò stretch of the SS121. The stretch has been subdivided into 22 homogeneous components of the groups 1a-1b-1c (embankments of the same height; embankments with supporting structures; simple trenches or sustaining walls; bridges) (Fig. 4). For each of the components a preliminary evaluation was carried out for structural vulnerability based on the criteria in Tables 1a–1d.

For the present application, in all the Tables 1a-1d a numeric value was assigned to each element considered in the evaluation (rows), increasing with the level of structural vulnerability. The values assigned in the tables were determined by executing a review of the literature (COOPER and VANDEPOL, 1991; BUCKLE and KIM, 1995; ZÜLFIKAR and YÜZÜGÜLLÜ, 1995; AIPCR, 1996; ARENA *et al.*, 1997; D'ANDREA, 1985; D'ANDREA and BOSURGI, 1999; AMBRASEYS and SRBULOV, 1995) using the same scale of level of vulnerability for each type of components. Then the structural vulnerability (Vs) of components was obtained by a sum of the values assumed by each element in the table. Analogously, an evaluation of functional factors G and Re was carried out based on all the road network (Tables 2a, 2b).

5. Evaluation of Seismic Risk and Analysis of Susceptibility

The aggregation (b) of the indicators hazard (*H*), exposure (*E*), vulnerability (*Vt*) allows the evaluation of seismic risk (*R*) of the sections of lifelines.

$$R = H^*E^*Vt. \tag{b}$$

This evaluation can be the basis of the planning of interventions of antiseismic retrofitting and definition of priority in relation to the available resources.

However, the evaluation of structural vulnerability of different components on the stretches of the network would need a long and onerous survey that would be difficult to manage and, in part, also useless. The generalization of the surveys also would involve elements characterized by negligible exposure, hazard, and functional vulnerability that would determine low levels of risk independently from the value of

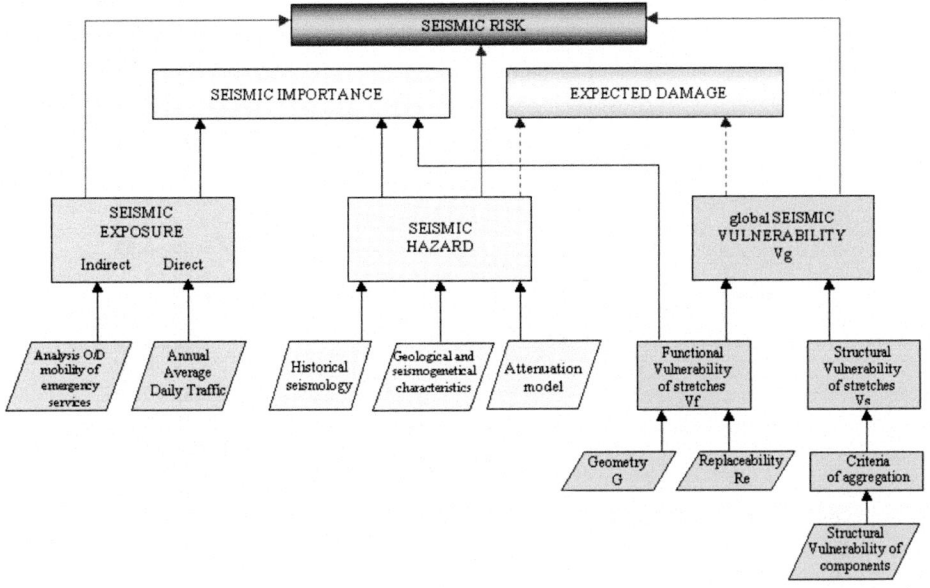

Figure 5
Flow chart of the procedure.

Seismic Risk on State Road n° 121

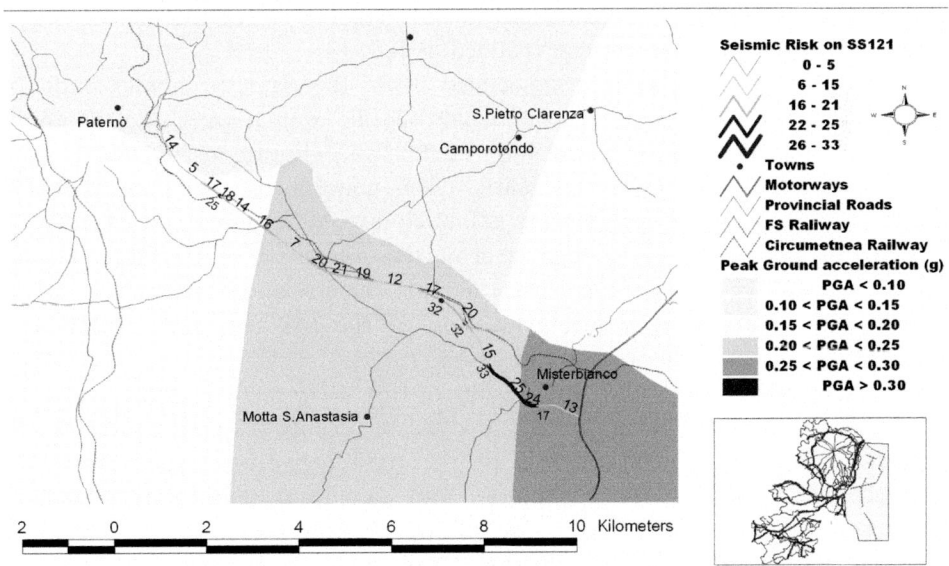

Figure 6
Seismic Risk of State Road 121.

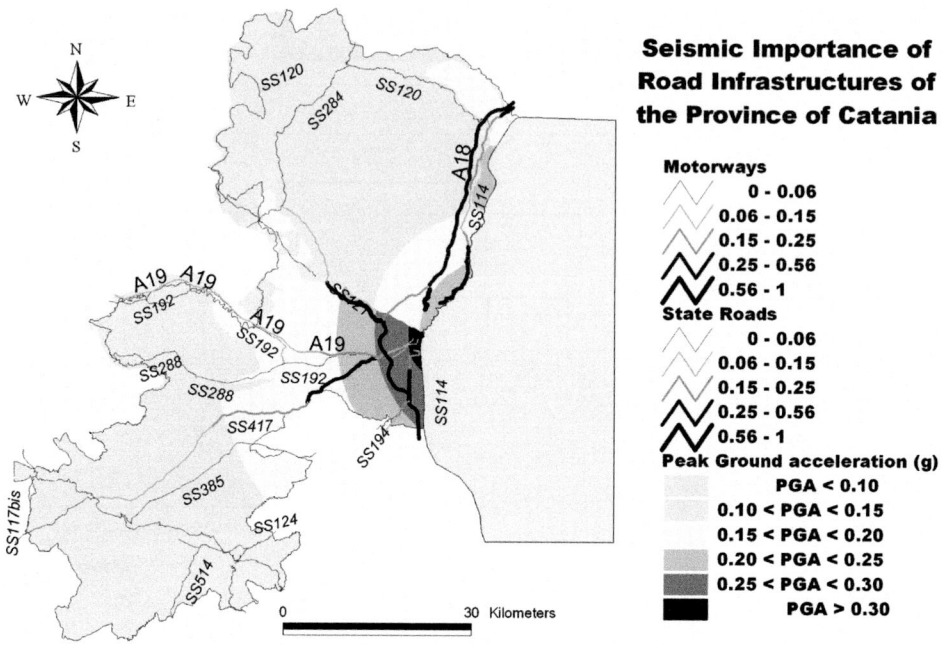

Figure 7
Seismic Susceptibility of road infrastructures of the Province of Catania.

structural vulnerability. This problem can be overcome by the use of "***analysis of seismic susceptibility***" that can be explained as follows.

The different stretches are associated with the same symbolic value of structural vulnerability ($Vs = 1$) without any specific evaluation. Using the above-mentioned procedure (b), it is possible to define an indicator of risk that assumes the meaning of "seismic susceptibility." The concept of *seismic susceptibility* applies to the whole network, with a limited use of resources, to identify the stretch that potentially has a high level of risk. On these routes more than on others, it would be necessary to closely identify the structural vulnerability of the components to arrive at a real evaluation of the risk.The total procedure is summed up in a flow chart (Fig. 5).

Figures 6 and 7 display the first results of this study relative to the evaluation of risk on state road 121 and to the analysis of seismic susceptibility on all the primary and secondary rural roads of the road network of the province of Catania. The analysis gave peremptory and significant indications relative to which are the more susceptible routes following the combination of the data on hazard, on the exposure and functional vulnerability: regarding all these routes, much more than on the others, it is important to determine the real seismic behavior of the components through studies of their structural vulnerability.

Figure 7 shows that one stretch of SS114 (between the bypass ramp and the junction with SS 194) has the maximum suceptibility value of the area under consideration, just because of its characteristics of irreplaceable necessity with in a context of maximum hazard.

6. Conclusion

In this study a methodology for the evaluation of seismic risk of rural road infrastructures is proposed. The methodology affirms the definition of the risk in relation to the level of hazard, exposure (direct and indirect), vulnerability (structural and functional) of each stretch of the network.

A preliminary phase of analysis of susceptibility is also proposed that provides, with a limited use of resources, the definition of the road lifelines that potentially have a high level of risk. The methodology, useful for the programming of intervention, for the reduction of the seismic risk of the road network and for the management of post-earthquake emergency, was applied experimentally in a restricted area of eastern Sicily and set up using opportune and representative indicators so that it will be possible to implement it with GIS software.

REFERENCES

AIPCR (Groupe de travail G2 AIPCR) (1996), *Natural Disaster Reduction for Roads – Comprehensive Report*, AIPCR - Association mondiale de la Route, World Road Association.
AMBRASEYS, N. and SRBULOV, M. (1995), *Earthquake-induced Displacements of Slopes*, Soil Dyn. Earthq. Engin. *14*, °(1).
ARENA, F., BOSURGI, G., and RICCIARDI, G. (1997), *Verifica probabilistica delle condizioni di resistenza di un rilevato soggetto ad azione sismica*, Atti 8° Convegno ANIDIS, Taormina 21–24 settembre 1997.
BUCKLE, I.G. and KIM, S. H. (1995), *A Vulnerability Assessment Model For Highway Bridges*, Lifeline Earthquake Engineering, Proc. Fourth U.S. Conference – Technical Council on Lifeline Earthquake Engineering, ASCE, Monograph No. 6 August, 1995.
CAFISO, S., CONDORELLI, A., and D'ANDREA, A. (1999), Evaluation of Seismic Risk on Road Infrastructures, PIARC XXI World Road Congress, Kuala Lumpur, 1999, Working group on Natural Disaster Reduction G2.
COOPER, T. R. and VANDEPOL, M. (1991), *Bridge Damage Review: Loma Pietra Earthquake*, Lifeline Earthquake, Technical Council on Lifeline Earthquake Engineering, ASCE, Monograph No. 49 August, 1991.
D'ANDREA, A. (1985), *Analisi delle condizioni limite di equilibrio dei muri di sostegno a gravità soggetti ad azioni sismiche*, Ingegneria sismica, anno II n°1 – Gennaio – Aprile 1985.
D'ANDREA, A. and BOSURGI, G. (1999), *Verifica delle condizioni di resistenza sismica di un rilevato stradale con gradonature laterali mediante analisi dinamica aleatoria*, Quarry and Construction, n°7.
D'ANDREA, A., CAFISO, S., ARGENTO, F., and EMMA, A. (1996), *Un S.I.T. per la gestione delle infrastrutture viarie della provincia di Enna*, Mondo GIS, 2.
D'ANDREA, A., LEONE, L., and MUSSUMECI, G. (1999), *Proposta di un S.I.T. per la valutazione del rischio sismico del sistema dei trasporti*, Congresso Nazionale S.I.F.ER.T., Parma, 1997.
PCM-DPS (Presidenza del Consiglio dei Ministri – Dipartimento della Protezione Civile), (1998), *Individuazione delle zone ad elevato rischio sismico del territorio nazionale*, Ordinanza n. 2788 del 12.06.1998 pubblicata su G.U. n.112 del 25.06.1998.

SABETTA, F. and PUGLIESE, A. (1996), *Estimation of Response Spectra and Simulation of Nonstationary Earthquake Ground Motions,* Bull. Seismol. Soc. Am. *86, 2.*

SABETTA, F. and PUGLIESE, A. (1987), *Attenuation of Peak Horizontal Acceleration and Velocity from Italian Strong-motion Records,* Bull. Seismol. Soc. Am. *77, 5.*

ZÜLFIKAR, A. C. and YÜZÜGÜLLÜ, Ö. (1995), *Preliminary Assessment of Seismic Vulnerability of Highway Bridges In Instanbul, Turkey,* Lifeline Earthquake Engineering. Proc. Fourth U.S. Conference, Technical Council on Lifeline Earthquake Engineering, ASCE, Monograph No. 6 August, 1995.

(Received July 7, 2001, revised December 9, 2002, accepted February 12, 2003)

 To access this journal online:
http://www.birkhauser.ch

GPSR Compliance

*The European Union's (EU) General Product Safety Regulation (GPSR)
is a set of rules that requires consumer products to be safe and our
obligations to ensure this.*

*If you have any concerns about our products, you can contact us on
ProductSafety@springernature.com*

In case Publisher is established outside the EU, the EU authorized
representative is:

Springer Nature Customer Service Center GmbH
Europaplatz 3
69115 Heidelberg, Germany

Batch number: 09491464

Printed by Printforce, the Netherlands